Rabbit Farming

NIPA® GENX ELECTRONIC RESOURCES & SOLUTIONS P. LTD.
New Delhi-110 034

About the Authors

Dr. Stephen Soren is an accomplished academic and professional specializing in Animal Nutrition. Currently, he is an Assistant Professor and Head of the Department of Animal Nutrition at West Bengal University of Animal and Fishery Sciences (WBUAFS) in Kolkata-700037. He earned his Master's degree (M.V.Sc) in Animal Nutrition in 2007 and worked as a Veterinary Officer at the Directorate of Animal Resources and Animal Health under the Government of West Bengal for nearly 8 years. Driven by a passion for teaching and research, Dr. Soren joined WBUAFS in 2018 as an Assistant Professor. He completed his PhD in 2024 and authored a book titled "Fodder Production and Principles of Animal Nutrition" in the same year. He has published 18 research articles in international and national journals and authored 6 book chapters. Dr. Soren is a life member of different professional societies like ANA & ANSI.

Dr. Amitava Roy is currently an Assistant Professor in Animal Nutrition at the Department of Livestock Farm Complex, West Bengal University of Animal and Fishery Sciences (WBUAFS), Kolkata, India. He obtained his Master's (M.V.Sc) in 2008 and Ph.D. in 2012 from Animal Nutrition division. Previously Dr. Roy engaged in Murshidabad Krishi Vigyan Kendra under W.B.U.A.F.S. as a Subject Matter Specialist for almost 7 1/2 years. He has published 36 mos. articles in International & National Journals, authored book chapters and has made contributions in different corners of Animal Nutrition. He has been involved in multidisciplinary research or field based projects to promote the scientific basis of Veterinary science for upliftment of livelihood. Dr. Roy is a life member of different professional societies like ISSGPU, APHV, ANA, IJBEAS & ANSI and also awarded PGDAEM, MANAGE.

Rabbit Farming

Stephen Soren, PhD
Assistant Professor and Head
Department of Animal Nutrition
West Bengal University of Animal and Fishery Sciences
Kolkata-700037, West Bengal, India

Amitava Roy, PhD
Assistant Professor (Animal Nutrition)
Department of Livestock Farm Complex
West Bengal University of Animal and Fishery Sciences
Kolkata-700037, West Bengal, India

NIPA® GENX ELECTRONIC RESOURCES & SOLUTIONS P. LTD.
New Delhi-110 034

NIPA® GENX ELECTRONIC RESOURCES & SOLUTIONS P. LTD.

101,103, Vikas Surya Plaza, CU Block
L.S.C.Market, Pitam Pura, New Delhi-110 034
Ph : +91 11 27341616, 27341717, 27341718
E-mail: newindiapublishingagency@gmail.com
www: www.nipabooks.com

For customer assistance, please contact
Phone: + 91-11-27 34 17 17
Fax: + 91-11-27 34 16 16

Print ISBN: 978-93-58878-33-2
ebook ISBN: 978-93-58878-58-5

Composed and Designed by NIPA®.

Preface

Rabbit farming has become increasingly popular in recent years, but many farmers or investors face challenges due to the lack of comprehensive guides and resources. To address this issue, "Rabbit Farming" offers a detailed and informative resource for rabbit farmers. This book covers breed selection, marketing strategies, health and nutrition, housing requirements, breeding techniques, and disease prevention and control. Whether you're new to rabbit farming or an experienced farmer, this book provides valuable insights and strategies for success in the industry. All images and figures in this book are sourced from the public domain to enhance reader understanding.

We would like to take this opportunity to express our deepest gratitude to all those who have played a crucial role in the making of this book. First and foremost, we extend our heartfelt thanks to the faculty members for their constant guidance, support, and encouragement. Their invaluable contributions have helped shape this book into its final form. We are also grateful to our colleagues and friends who have provided us with valuable insights, feedback, and motivation throughout this journey. We would also like to acknowledge the unwavering dedication and hard work of the supporting staffs who have assisted us in various stages of this project. Special thanks are due to Nipa Genx Electronic Resources and Solutions Pvt. Ltd., New Delhi, for their expert publishing services that have helped bring this book to fruition. Lastly, we cannot forget the unwavering support and understanding of our family members, whose love and encouragement kept us going during the writing process.

We sincerely hope that this book fulfils its objectives and becomes a valuable resource for students, researchers, faculty, and all interested individuals seeking knowledge in this field. It is because of all your support that we were able to complete this book, and for that, we are truly grateful.

Stephen Soren
Amitava Roy

Contents

Preface ... v

1. **Overview of Rabbit Production** ... **1**

2. **History of Rabbit Production** ... **11**

3. **Common Breeds of Rabbits** ... **13**

A. Rabbit Breed For Wool Production ... 13

Benefits of Raising Angora Rabbits ... 19

Raising Angora Rabbits ... 19

B. Rabbit Breed For Meat Skin Production ... 20

C. Rabbit Breed For Fancy/Hobby Purposes ... 29

4. **Housing and Management Practices** ... **35**

Wire Hutches ... 35

Wood-frame and Wire Hutches ... 38

Bamboo Hutch ... 40

Hutch Floor ... 40

Feeding Equipments ... 41

Types of Feeders ... 42

1. Crocks ... 42

2. Grass Mangers ... 43

3. Bamboo Troughs ... 43

4. Hoppers ... 44

Waters ... 47

Types of Water Containers ... 47

Rabbit Housing Options for Optimal Care and Management: ... 50

Cage System ... 50

Hanging Cage System ... 51

Hutch System ... 52

Colony Housing System ... 53

Rabbit Tractor Cages ... 54

Nest Boxes 55
1. Standard Nest Box 55
2. Counterset Nest Box 57

5. Feeding and Nutrition Requirements....................59
Feeding of Rabbits 59
Water Requirements....................61
Carbohydrates for Rabbits....................62
Recommended Fiber Levels for Rabbits....................62
Concentrates....................63
Home Mixed Feed....................65
Feeds for Feeding Rabbits....................66
Don't Feed: 66
DO Feed: 67
Coprophagy....................67
Lactation....................68

6. Management of the Herd....................69
Materials Needed 69
Guidelines 70
Weaning Guidelines....................70
Ways to Care for Rabbits....................72
Using Palpation to Assess Pregnancy 73
Kindling 75
Young Litter Care 76
Reasons for Losses in Infant Litters....................77
1. Loss of milk 77
2. Eating young....................77
3. Disturbance 77
4. Weaning....................78
Determining the Rabbits' Gender....................78
Record Keeping....................79
Preventing Accidents 83
Hygiene and Disease Control....................84

Prevention of Diseases 85
Other recommendations for illness prevention are as follows: 87
Habit of 'Fur' Eating 88
Avoiding 'Fur' Block 88
Gnawing the Hutch's Wooden Parts 89
Ideal Environment for Rabbit Rearing: 89

7. Management of Rabbit Farm Reproduction 93
1. Breeds and Selection 94
2. Rearing future does 95
3. Mating 96
4. Artificial Insemination (AI) 97
5. Hormonal treatment and biostimulation 99
6. Lighting program 99
7. Doe-litter separation (DLS) 100
7. Suckling mortality 101
8. Teat number 102
9. Seasonal effect 102

8. Common Diseases of Rabbits and Control Measures 105
1. Snuffles in Rabbits 105
2. Cutaneous Staphylococcosis 106
3. Abscesses under the skin 107
4. E. coli Infections 108
5. Wry neck condition in Rabbits 109
6. Coccidiosis 109
7. Sore Hock 110
8. Caecal Impaction 111
9. External Parasites 112
10. Hairballs / Wool Blocks 112
11. Myxomatosis 113
12. Toxoplasmosis 114
13. Ring Worm 114
14. Encephalitozoonosis 115

15. Rabbit Syphilis 115
16. Moist Dermatitis 115
17. Tyzzer's Diseases 116
18. Red Urine 116
19. Indirect Antibiotic Toxicity 118
Preventing general illnesses in rabbits 118
Cleanliness 118
Ventilation 118
Observation 118

9. Marketing and Profits 119
Rabbit Meat Production 119
Consumer Preference and Demand for Rabbit Meat 123
Consumer Attitudes Towards Functional Foods Including Rabbit Meat 126
Rabbit Farming Project (100+20) Cost and Profits 127
Introduction of the Project 127
Rabbit Breeding & Rearing Cycle in Rabbit Farming Project 128
Pregnant Rabbit 128
Newborn Rabbits 129
Capital cost/expenditure for Rabbit Farming Project 130

10. Future Rabbit Research in India 133
Breeding and Genetic Research 133
Meat Science Research 134
Wool Science Research 135
Reproduction Research 135
Health Management Research 136
Nutrition Research 136
Physiology and Climatology 137
Social Science Studies 137
Organic Rabbit Farming 138

Index 139

1

Overview of Rabbit Production

Food security is a pressing global concern that requires immediate attention. The rapid pace of urbanization and industrialization has resulted in a reduction of available land for agriculture, making it increasingly difficult to meet the growing demand for food. In countries like India, where there is already a shortage of meat, rabbit farming has shown great potential in helping to bridge this gap.

The rabbit farming industry has experienced significant growth as traditional gender biases have diminished, attracting a wider range of individuals to get involved. Efforts to promote the nutritional value of rabbit meat are gaining momentum, with the aim of integrating it into the mainstream market. This could have positive implications for both producers and consumers alike. With the rise of industrialized rabbit production units – similar to those used in poultry farming – modern and efficient methods are now being utilized to increase productivity and profitability. This shift not only meets consumer preferences for sustainable and high-quality food but also provides smaller-scale producers with an opportunity to compete on a level playing field with larger operations. The practice of rabbit husbandry, while relatively new, has the potential to make a significant impact, particularly in agricultural regions. The meat of rabbits is highly sought after for its palatability and nutritional value, without any religious or cultural limitations. Moreover, the demand for rabbit wool, especially from breeds such as German Angora, continues to rise due to its efficient production. By incorporating rabbit farming into existing agricultural systems, farmers can diversify their sources of income and improve food security in their communities.

Rabbit meat is an excellent source of affordable protein, making it particularly beneficial for communities facing food insecurity. Its availability and affordability make it an ideal option for addressing malnutrition in developing regions. Additionally, by using all parts of the rabbit - from its meat to its fur and even manure - farmers can take advantage of additional economic opportunities. With advancements in technology and a growing trend towards organic produce, the rabbit farming industry is constantly evolving and thriving within the realm of agriculture. The future looks promising with increased support from governments and organizations focused on achieving sustainable development goals. Rabbit husbandry has already shown its potential to significantly contribute towards promoting sustainable agriculture practices and improving overall livelihoods in rural communities.

Characteristics of rabbits that make them suitable as meat-producing micro livestock in developing countries like India are as follows:

1. **Small body size:** Small body size is a key characteristic of rabbits that makes them unique and versatile in many ways. Due to their small size, they require only a small amount of feed, making them cost-effective for owners. In addition, their houses can be easily constructed with inexpensive materials. This not only makes it affordable but also convenient for owners to provide a suitable living space for their rabbits. Furthermore, the small body size results in a smaller carcass which can be consumed by a family in one meal. This eliminates the need for storage, making rabbits an ideal option as a "biological refrigerator". The Soviet Chinchilla breed has a mature body weight ranging from 3.5 to 4.0 kg while the New Zealand White breed varies from 4.0 to 5.0 kg. Similarly, the Gray Giant breed has a weight range of 4.0 to 4.5 kg and the White Giant breed ranges from 4.5 to 5.5 kg when fully matured. This makes these breeds popular choices among farmers and households looking for an efficient source of food production and consumption.
2. **Short generation interval and high reproductive potentiality**: Rabbits have a short generation interval and high reproductive potentiality, making them an ideal animal for breeding purposes. Unlike some other animals, rabbits are induced adulators which mean they are able to breed within 24 hours of giving birth. Additionally, their gestation period is only one month, making it quite feasible to obtain 5-6 crops in a year from just one doe. On average, rabbits produce 6-7 kits per litter, which means that in just one year, a single doe can produce an impressive number of 30-40 kits. This is especially true for breeds such as SC, NZW, CW etc., which are known for their ability to produce even higher numbers of kits ranging from 35-40 per litter. This makes rabbits an extremely efficient and highly desirable animal for breeding purposes.
3. **Potential for genetic improvement**: The potential for genetic improvement in rabbits is highly promising due to the wide range of diversity within their genetic resource pool. With mature body weights varying from 2-6 kg and a multitude of traits such as maternal ability, fecundity, and resistance to heat stress exhibiting high levels of variability,

rapid improvements in performance can be achieved through selective breeding using different breeds. Not only are rabbits genetically flexible, making them adaptable to various production systems, but their short generation interval allows for a quick selection process. Additionally, the heritability of growth and carcass traits are moderately high, allowing for swift advancements in these areas. This presents great opportunities for producers looking to improve the overall quality and productivity of their rabbit populations through strategic breeding techniques.

4. **Rapid growth:** Rapid growth is a key characteristic of rabbits, with an average daily gain that ranges from 20-40g/day in temperate regions, depending on the breed. This remarkable rate of growth is unmatched by any other animal, aside from poultry. Studies have shown that the quick growth and high prolificacy of rabbits contribute to their impressive meat production capabilities, with a doe able to produce more meat per unit live weight per year compared to other livestock animals. These factors make rabbits an attractive option for meat production, particularly in areas with limited space or resources, as they can reach consumable size at a fast rate while utilizing less land and feed. Rapid growth is truly one of the many strengths of these small but mighty animals.
5. **Good mothering ability**: The good mothering ability of rabbits is evident in their high reproductive rate, with a number of young per doe per year that is higher than most animals. In fact, rabbits have a higher reproduction rate compared to all animals except fowls and turkeys. This showcases the remarkable maternal instincts and nurturing qualities possessed by these small mammals. With proper care and management, rabbits can produce multiple litters each year, making them an efficient source of meat and fur. The strong bond between rabbit mothers and their offspring ensures the survival and growth of their young, highlighting their excellent mothering abilities. This exceptional trait makes rabbits popular among farmers and pet owners alike, as they provide both practical benefits and endearing companionship.
6. **Utilization of non-competitive food:** Rabbits are a unique animal that falls between ruminants and non-ruminants, making them a valuable option in terms of food utilization. Unlike pigs and poultry, rabbits do not compete with human food, making them an ideal choice for sustainable farming. Similar to ruminants, rabbits can thrive on a diet with as little as 20% grains, making their rearing more cost-effective. Producing one kilogram of body weight gain in rabbits only requires around 3.5 kilograms of feed. This efficient conversion of cellulose-rich

roughage into meat makes rabbits a valuable source of food in countries facing severe food shortages. Moreover, unlike pigs and fowl, rabbits can be solely maintained on roughage diets, reducing feed costs by 50%. In comparison, the feed cost for cattle and pigs can add up to 65-70% of the total rearing cost. Therefore, the utilization of non-competitive food sources such as rabbits can have significant benefits for both farmers and areas with limited access to sufficient food resources.

7. **Most efficient meat producing animal:** Rabbits are recognized for their exceptional meat production efficiency, attributed to their unique behavior known as caecotrophy. This behavior enables them to consume their caecal content, enhancing protein extraction from their diet. Research has demonstrated that rabbits only need 105 Kcal of energy to produce one gram of protein, a significantly lower requirement compared to pigs, sheep, beef cattle, broiler chickens, and turkeys. It is noteworthy that producing one kilogram of rabbit meat necessitates only one-quarter of the feed energy required for beef production. Additionally, rabbits exhibit lower energy and protein needs per unit of live weight gain in comparison to other livestock species. These findings underscore the superior energy and protein utilization efficiency of rabbits as meat producers relative to other livestock. Given their efficient feed-to-meat conversion, rabbits play a crucial role in the meat industry and can substantially contribute to meeting the escalating demand for animal protein.
8. **Good carcass quality:** Rabbit meat is known for its good carcass quality, making it stand out among other animal species. It is rich in protein, essential minerals like calcium and phosphorous, and certain vitamins such as nicotinic acid and calcium pantothenate. However, it contains less energy and fat compared to other meats. Furthermore, rabbit fat has a higher proportion of polyunsaturated linolenic and lonoleic acid which makes it suitable for individuals with cardiovascular issues or those who are aged. The dressing percentage of rabbit meat typically ranges from 55-65%, not including the weight of the head. However, some communities also consume the head, bringing the dressing percentage up to an impressive 62-72%. Apart from its nutritional value, rabbit meat also boasts high organoleptic properties such as tenderness, juiciness, and flavor. Its high moisture content and water holding capacity contribute to its desirable juiciness when cooked properly. Overall, rabbit meat offers a superior carcass quality that sets it apart from other types of meat.

9. **Production of fur skin**: The production of fur skin is a significant by-product of the rabbit industry, in addition to its meat and wool. Once the rabbits are slaughtered, their pelt undergoes processing to be transformed into various products such as jackets, caps, bags, gloves, and more. The larger sized fur skins and those with a white color tend to be more highly valued due to their advantage in being easily dyed. There are several dual-purpose breeds of rabbits that are specifically bred for both their meat and fur skin production, including Soviet Chinchilla, New Zealand White, Gray Giant, White Giant, and California White. This thriving industry not only provides a valuable source of income for farmers but also offers consumers high-quality and sustainable fur products.
10. **Production of high quality wool**: Angora wool is a highly valued and sought-after product in the international wool trade. It is known for its exceptional quality and is used in the production of various woolen garments such as shawls, sweaters, and blankets. What sets Angora wool apart from other types of wool is its low friction coefficient, which gives it a soft and luxurious feel. Additionally, due to its long and medullated fibers, Angora wool is much lighter than traditional wool. On average, an adult rabbit can produce 750-900g of Angora wool per year, with a market value of around Rs. 900/- per kilogram. There are various breeds of rabbits that are specifically raised for their high-quality Angora wool, such as the German Angora, Russian Angora, and British Angora. Interestingly, while Angora wool is the main product from these rabbits, their meat also serves as a byproduct and is available under the name "New Zealand White (NZW)" during slaughter. Overall, the production of high-quality Angora wool plays a significant role in the global market and continues to be favored for its exceptional properties.
11. **Rabbit manure:** Rabbit manure, also known as bunny berries, is a highly sought-after source of fertilizer for vegetable growers. It is rich in essential nutrients such as nitrogen (2.91%), phosphorous (1.94%), and potassium (1.70%), making it a valuable asset for improving soil quality and promoting plant growth. In fact, the nutrient levels in rabbit manure are higher compared to other types of animal manure, making it a popular choice among farmers and gardeners worldwide. It is estimated that 50 adult rabbits can produce anywhere between 5 to 7 kilograms of manure per day, making it a reliable and sustainable source of organic fertilizer. Additionally, rabbit manure is odorless and does not attract pests or burn plants, making it easy to handle and use in various gardening practices. With its high demand and benefits for crop production, rabbit manure continues to be valued by agricultural communities globally.

12. **Easy management:** Easy management is one of the key advantages of raising rabbits as pets or for commercial purposes. Unlike other livestock, the care and maintenance required for rabbits is minimal, making it suitable for people of all ages and physical abilities. Whether it is an aged person, a small boy, or a lady, anyone can manage a rabbit farm with ease. Furthermore, the labor requirement is also significantly low - only one laborer is sufficient to manage 150 rabbits. This not only reduces the cost of labor but also makes it accessible to individuals with limited resources. The task of managing rabbits simply involves ensuring their living space is clean and providing them with food on a regular basis. With proper planning and organization, managing a rabbit farm can be effortless and hassle-free while still yielding high profitable output.
13. **No religious taboo**: In today's diverse society, where people of various religions and beliefs coexist, it is important to understand and respect each other's cultural norms. When it comes to food choices, religious sentiments often play a significant role in determining what one can or cannot consume. However, when it comes to rabbit meat, there appears to be no religious taboo against its consumption among the non-vegetarian population. Unlike some other meats that may be prohibited in certain religions, such as pork in Islam and beef in Hinduism, there is no specific mention of rabbits being forbidden in any major religion. In fact, in Christianity and Judaism, rabbit meat is considered a clean and permissible food source. This lack of religious restriction allows for the widespread consumption of rabbit meat without any hesitation or conflict with one's faith. Therefore, it can be safely concluded that unlike some other types of meat, there is no religious sentiment against consuming rabbit meat among the non-vegetarian population.
14. **Less disease prone:** Rabbits are known to be less disease-prone compared to poultry, making them a more reliable and sustainable source of meat production. Unlike poultry, which are susceptible to diseases such as RD, GD, fowl cholera, and fowl pox that can decimate entire populations, rabbits have a lower susceptibility to fatal diseases. The only significant threat to rabbits is coccidiosis, which can be effectively managed through regular deworming on a monthly basis. This resilience makes rabbits an ideal choice for meat production, as their risk of disease outbreaks is significantly lower than that of other livestock. This not only ensures the safety and health of the animals but also leads to a more consistent and higher output for farmers.

Although rabbit production offers numerous benefits for sustainable livelihood development, it is essential to understand the potential challenges before embarking on this farming endeavour. Therefore, farmers must conduct comprehensive research and carefully evaluate these constraints to guarantee a successful and sustainable rabbit production operation. The primary constraints of rabbit farming are outlined below:

1. **Climate:** The climate plays a crucial role in the life and well-being of animals, especially those living in the wild. In particular, rabbits face various challenges due to their lack of functional sweat glands and poor thermoregulatory mechanism. This makes them highly susceptible to extreme temperatures, where an ambient temperature of 1000 F (37.8 C) can be fatal for them due to discomfort and overheating. The ideal temperature range for rabbits is between 5°C to 32°C, with a relative humidity of 60 to 80 percent. However, this can be a major constraint in rabbit production as it requires careful management and control of the temperature through insulation methods in housing construction, particularly in large-scale rabbitries. By providing insulated materials for housing, we can ensure a suitable environment for rabbits to thrive and produce efficiently. Therefore, understanding and addressing the impact of climate on rabbits is vital for their welfare and successful production.
2. **Cages:** The construction of pucca houses for raising rabbits requires a significant investment in large scale rabbiteries. However, for small scale rabbit farming, alternative options such as raising them inside buildings or outside in hutches with protective roofs or cages can also be utilized. These cages can be constructed using locally available materials such as bamboo splits or wire mesh. This allows for a budget-friendly and sustainable approach to rabbit farming, while still ensuring the safety and protection of the animals. It also provides flexibility in terms of space utilization and can easily be expanded as the need arises. By utilizing these alternatives, rabbit farmers can effectively manage their costs while achieving optimal output in their production process.
3. **Rabbit feed:** Rabbit feed is an essential aspect of raising healthy and productive rabbits, whether in small backyard setups or large-scale rabbitries. In small-scale or backyard rearing, rabbits can be fed with vegetable kitchen wastes or fresh greens, supplemented with small amounts of concentrates. This feeding option is not only cost-effective but also provides a varied diet for the rabbits. However, when it comes to large-scale rabbitries, relying solely on such feeding methods may

not be feasible due to the volume of rabbits being raised. In such cases, pelleted feeding is often used as it provides a complete and balanced nutrition for the rabbits. While this option may be more expensive, it ensures that the rabbits receive all the necessary nutrients for optimal growth and productivity. Therefore, choosing the right type of rabbit feed based on scale and resources is crucial in ensuring the health and success of a rabbitry operation.

4. **Marketing:** Marketing plays a crucial role in promoting any product or service, and the same applies to rabbit meat. Currently, there is a lack of a well-defined market for rabbits and rabbit products in the country. This presents a major challenge for breeders and farmers who are looking to expand their business in this industry. Therefore, it is imperative that development agencies give due attention to the marketing of rabbit meat in order to popularize rabbit breeding on a larger scale within the country. With effective marketing strategies and campaigns, the demand for rabbit meat can be increased, leading to growth opportunities for breeders and farmers. This would not only benefit the agricultural sector but also provide consumers with an alternative source of protein. Thus, proper marketing efforts can have a significant impact on the success of rabbit breeding in the country.
5. **Rabbit research/development:** The research and development of rabbit production for meat is a crucial aspect for the planned and scientific rearing of domestic rabbits. In order to ensure the success of this industry, it is important for concerned agencies in the country to strengthen their efforts in research, development, extension, and training programs. These initiatives can help improve breeding techniques, optimize feed formulation, and enhance disease management strategies. Additionally, conducting research on the genetic selection and improvement of rabbits can lead to higher quality meat production. The dissemination of these findings through extension programs and training workshops can also educate farmers on best practices for raising rabbits as livestock. By investing in rabbit research and development, we can improve the overall productivity and profitability of this industry while meeting the growing demand for sustainable protein sources.
6. **Disease susceptibility**: Disease susceptibility is a significant concern in rabbit farms as it can have a detrimental impact on production levels, ultimately leading to unprofitable outcomes. Rabbits are highly susceptible to various diseases, making them vulnerable to respiratory infections caused by *Pasturella multocida*, tuberculosis, and pseudo-

tuberculosis. Along with these, they are also prone to enteric diseases such as enterotoxaemia and ceacal impaction. The threat of viral infections like myxomatosis and rabbit pox only adds to the already high morbidity and mortality rates in rabbits. To make things worse, parasitic diseases such as Coccidiosis, tanaesis, and mange are also common among rabbits. Thus, it is imperative for rabbit farmers to implement strict management and prevention measures in order to maintain a healthy population of rabbits and sustain their productivity levels.

We cannot underestimate the importance of integrating rabbit farming into existing agricultural systems. Not only does it provides various economic opportunities but also plays a crucial role in enhancing food security globally. With continued support and advancements in technology, we can expect to see further growth and success within this industry as we strive towards a more sustainable future.

2

History of Rabbit Production

Rabbits belong to the mammalian order Lagomorpha, which includes rabbits, hares, and Pikas. Fossil records indicate that Lagomorpha originated in Asia over 40 million years ago during the Eocene period. The continental drift during this period likely contributed to the diverse distribution of rabbit and hare species worldwide, excluding Australia. There are over 60 recognized breeds of domestic rabbits in Europe and America, all descended from the European rabbit (Oryctolagus cuniculus), the only widely domesticated rabbit species. This species is distinct from native rabbits like North American jackrabbits and cottontail rabbits, as well as all hare species.

The European wild rabbit evolved approximately 4,000 years ago on the Iberian Peninsula, where the Phoenician merchants named the region 'Hispania,' meaning 'land of the rabbits.' When the Romans arrived in Spain around 200 BC, they began farming native rabbits for meat and fur, a practice known as 'cuniculture.' The Latin name 'Oryctolagus cuniculus' translates to 'hare-like digger of underground tunnels,' reflecting the rabbits' behavior. The Roman Empire's expansion and increased trade facilitated the introduction of the European rabbit to various parts of Europe and Asia. With their rapid reproduction and suitable habitat, wild rabbit populations thrived in new locations, including North America and Australia, where they became agricultural pests.

Monks in the Champagne Region of France are believed to have first domesticated rabbits in the 5th century for food. Rabbits were introduced to Britain in the 12th century, leading to widespread breeding and farming for meat and fur during the Middle Ages. Selective breeding efforts resulted in distinct rabbit breeds emerging in different regions, with some breeds tracing back several centuries. By the 19th century, a wide variety of rabbit breeds had been developed, ranging from small Polish rabbits to large Flemish Giants.

During the Victorian era, new 'fancy' breeds were created for rabbit shows, and rabbits became popular pets among the middle class. The romanticized view of rabbits as pets persisted into the 20th century, leading to the development of new breeds and colors through selective breeding and cross-breeding. Rabbit breeding became a popular hobby in Europe, with breed societies and clubs established to promote different breeds. In the United States, domestic rabbitry gained popularity in the early 20th century, with European breeds being imported and American breeds developed.

During the World Wars, governments across the world recognized the potential of rabbits as a solution to food shortages. Thanks to their rapid reproduction and efficient conversion of feed to meat, rabbits earned the nickname "Underground Mutton" and became a vital source of sustenance for many rural communities. Even after the wars ended, rabbits remained a popular option for both food and fur production. However, in recent years, attitudes towards rabbits have undergone a significant shift. Instead of being viewed solely as a source of food or fur, they are now valued for their intelligence, unique personalities, and playful behavior. In fact, in the United Kingdom, rabbits rank third as the most popular pet. Owners now prioritize the welfare of their furry companions and provide them with proper healthcare and regular interaction. Rabbits are no longer seen as mere livestock but instead are treated with the same level of care and affection typically reserved for cats and dogs. These changes in attitudes towards rabbits reflect an increased understanding of their needs and behaviors. As valuable companions, they play an integral role in many households around the world today.

In India, the practice of raising rabbits for commercial purposes gained significant traction in the early 1960s with the introduction of the German Angora breed in regions such as Himachal Pradesh and Jammu and Kashmir. This breed was known for its high-quality wool production, making it a lucrative choice for farmers. Recognizing the potential of this industry, the Indian government took further steps to support it by establishing a Central Research Station in 1977. This institution aimed to assist rabbit farmers in overcoming challenges related to breeding, disease prevention, and promoting sustainable practices.

Over the years, rabbit farming has spread across various states in India, catering to both wool production and an emerging broiler market. Regions with temperate and subtropical climates like West Bengal, Assam, Manipur, Andhra Pradesh, Tamil Nadu, Kerala,and Karnataka have seen significant growth in this sector due to their favorable weather conditions. Today, rabbit farming is not only a source of income for many farmers but also contributes to India's overall economy through its export of quality wool and meat products. With continued support from the government and advancements in technology and breeding techniques, the future looks bright for this industry in India.

With advancements in technology and growing demand for organic meat products worldwide, rabbit farming continues to evolve and thrive as an important sector within agriculture. The future looks promising for this industry with increased support from governments and organizations working towards sustainable development goals.

3

Common Breeds of Rabbits

Breeds of rabbits: There is a diverse range of breeds when it comes to domestic rabbits, with each one possessing unique qualities. Currently, there are 38 recognized breeds and 87 varieties of rabbits that are well established all around the world. These breeds and varieties differ in various aspects such as color, size, type of hair coat, and other characteristics. Depending on their purpose, rabbits can be categorized into three groups: those bred for meat production, those bred for fur production, and those kept as companion animals. Each breed has its own distinct traits and characteristics that make them suitable for their designated purpose. Whether you're looking for a fluffy pet or a high-quality fur producer, there is a breed of rabbit that will fit your needs. With the large number of breeds available, there is no shortage of options when it comes to choosing the perfect rabbit companion.

A. Rabbit Breed For Wool Production

There are several rabbit breeds known for their wool production. These breeds have been selectively bred for their high-quality wool. They provide a sustainable and eco-friendly alternative to traditional wool sources, making them valuable in the textile industry.

Angora rabbit: The Angora rabbit is a breed that is believed to have originated in Ankara, Turkey, although the exact origins of this fluffy and highly coveted animal remain unclear. What we do know for certain is that Europe has been raising angora rabbits for their luxurious fiber for centuries, with the French being credited for making their wool popular around 1790. However, it wasn't until 1920 that North America would see the introduction of this luxury fiber. The soft and silky wool produced by these rabbits has been highly sought after for its warmth and softness, making it a valuable commodity in the textile industry. The angora rabbit's popularity continues to grow worldwide as more people discover and appreciate the beauty and utility of this fascinating breed.

There are several strains and breeds of Angora rabbits commonly raised by farms in India. These rabbits have been selectively bred for their high quality wool, among other desirable traits.

1. German Angora: German Angora, originating from Germany, is a highly sought after breed known for its exceptional wool production. This breed typically produces 700-1000 grams of luscious wool per year, with a soft and fine white texture. The wool is highly valued in the garment industry due to its warmth, softness, and durability. German Angoras have an adult body weight of 3-4 kilograms and are known for their docile nature and easy handling. They have a distinctive coat with 2-4% guard hair that helps protect them from harsh weather conditions. With their high wool yield and quality, German Angoras are a preferred choice among breeders and farmers worldwide.

Fig. 1: German Angora

2. British Angora: British Angora rabbits, originating from the United Kingdom, are known for their beautiful wool production. While they may produce slightly less wool compared to other angora breeds, with an average of 400-600 grams per year, their fine quality and low percentage of guard hair (2-4%) make up for it. These rabbits have an adult body weight ranging from 2.5-4.5 kilograms, making them a manageable size for handling and grooming. In addition to their soft and lustrous fur, British Angora rabbits make gentle and sociable companions, making them a popular choice among rabbit owners. Their origin from the U.K. also adds a unique touch to these charming creatures, solidifying their place in the world of Angora rabbit breeding.

Fig. 2: British Angora

3. Russian Angora: Russian Angora is a breed of domestic rabbits originally from Russia. They are known for their luxurious and soft white wool, which yields around 300-400 grams per year. This wool has a medium fineness, with the added characteristic of having a higher percentage (10-20%) of guard hair. Due to this, the wool produced by Russian Angoras is more durable and suitable for various uses such as knitting and weaving. These rabbits are larger in size compared to other angora breeds, with an adult body weight ranging from 3.5 to 5.5 kilograms. Apart from their valuable wool production, Russian Angoras also make great pets due to their docile nature and charming personalities. With its combination of size, yield, and quality of wool, the Russian Angora remains a popular choice among rabbit enthusiasts and fiber artists alike.

Fig. 3: Russian Angora

4. French Angora: The French Angora is a breed known for its dense undercoat and low maintenance compared to other Angora breeds. Small ear tufts are allowed but not preferred by breeders. The American Rabbit Breeders Association (ARBA) recognizes the same colors as the English Angora, with the addition of ticked and wide band. French Angoras are categorized as 'white' or 'colored' at ARBA shows, with broken considered a colored variety. Toenails should be one color as per ARBA guidelines. This breed is larger, weighing between 3.4-4.8 kilograms, with a commercial body type. It has a clean face and front feet, minimal tufting on the rear legs, and the color is determined by the head, feet, and tail matching. The wool is silky smooth, which can be challenging for beginner spinners. Desirable characteristics of the fiber include texture, warmth, light weight, and pure white color. It is used for sweaters, mittens, baby clothes, and millinery.

Fig. 4: French Angora

5. Giant Angora: The Giant Angora is the largest among the ARBA-recognized Angora breeds. It was originally bred to be a productive commercial producer that could thrive on a diet of 16–18% protein pellets and hay, housed in standard-sized, all-wire cages. Due to ARBA restrictions on showing German Angoras, Louise Walsh of Taunton, Massachusetts, developed the Giant Angora by crossbreeding German Angoras, French Lops, and Flemish Giants to create a distinct 'commercial' body type. The breed was officially recognized by ARBA in 1988. The Giant Angora's coat consists of three types of wool: soft underwool, awn fluff, and awn hair. The awn-type wool is unique to the Giant and German Angora breeds. The Giant Angora features facial and ear furnishings, distinguishing it from the German Angora. The only color variety recognized by ARBA for the Giant Angora is the Ruby-eyed White (REW), indicating a genetic absence of pigment.

The Giant Angora produces more wool than other Angora breeds and is the only 6-class animal in the Angora breed. It should have a dense wool coat and a commercial-type body. The head should be oval-shaped, broad across the forehead, and slightly narrower at the muzzle, with noticeable forehead tufts and cheek furnishings.

The coat of the Giant Angora contains three fiber types: underwool, awn fluff, and awn hair. The under wool should be the dominant type, soft, medium-fine, and delicately waved. The Awn Fluff has a guard hair tip and is stronger and wavy, while the Awn Hair is a straight, strong hair that protrudes above the wool. The classification of the Giant Angora differs from other breeds due to its 6-class system. Judges evaluate the wool density, texture, and length, as well as the overall body type, head, ears, eyes, feet, legs, and tail. Giant

Angoras grow slowly, with does taking over a year to reach maturity and bucks up to 1.5 years.

Fig. 5: Giant Angora

6. Satin Angora: The Satin Angora breed, renowned for its unique sheen and rich coloration, was developed in the late 1970s by Mrs. Meyer from Canada. Through crossbreeding French Angoras with Satin rabbits, Mrs. Meyer successfully introduced red and copper pigments into the rabbits' coats, resulting in a strikingly beautiful appearance. The satinized hair shafts of this breed have a semi-transparent outer shell that reflects light, giving the fur a deep color, high lustre, and an incredibly soft and silky texture. Unlike the French Angora, the Satin Angora lacks facial, ear, or foot furnishings but is currently being selectively bred to increase wool production. While grooming can be more demanding than other Angora breeds like the Giant or French Angora, it is easier compared to the English Angora. A daily combing routine is recommended to prevent matting due to its soft wool texture and lower guard-hair count. Satin Angora wool is known for its strength in spinning, but its slippery nature can make spinning more challenging compared to other Angora varieties.

Fig. 6: Satin Angora

7. Crossbred Angora: Crossbred Angoras have been developed specifically for the Indian climate and conditions, with a wool yield of approximately 500-600 grams per year, making them ideal for small-scale production. They have white medium-fine wool with 4-8% guard hair, making their fleece soft and durable. These rabbits maintain an average adult body weight range of 3-5 kilograms, making them easier to handle compared to larger purebred Angoras. This makes them a suitable choice for farmers or breeders looking to diversify their breeding stock while still obtaining high-quality wool.

Fig. 7: Crossbred Angora

About Angora rabbit Wool

Angora rabbit wool is highly sought after and regarded as a premium fiber in the textile industry. It can be sold in its raw form, spun, dyed, or left natural. Due to its exceptionally fine texture, Angora wool is often blended with other fibers like sheep's wool, mohair, silk, and cashmere. The delicate nature of Angora wool makes it unsuitable for dense knitting stitches.

Angora wool is known to be seven times warmer than sheep's wool, making it too insulating for direct use in garments. However, when blended with other fibers, it adds softness, warmth, and a distinctive "halo" effect to the resulting yarn and garment.

Harvesting Angora Wool

Angora wool is obtained from rabbits through either plucking or shearing. Some breeds, like the English Angora, naturally shed their wool three to four times a year, a process known as molting. The timing of molting can vary depending on the breed and the specific lineage. The initial harvest typically occurs when the rabbits are between 5 and 8 months old, with Giants being slightly older and Satins younger. Signs that the rabbits are ready for harvesting include wool trailing behind them, scattered in their cages, and coming off in larger quantities during grooming. Subsequent harvests should be done every four months to align with the rabbit's natural molting cycle.

Breeders with rabbits that molt naturally can take advantage of this by gently plucking the loosened fiber during the molting period. Alternatively, wool can be harvested using scissors or clippers. Plucked wool generally fetches a higher price than cut wool. When cutting, ensure that the shears are sharp and trim small sections at a time to avoid harming the rabbit. The method of harvesting, whether plucking or cutting, should be chosen based on the breed. For instance, plucking a German Angora would be harmful as it may pull out the wool from the roots, while clipping French and English Angoras could lead to matting due to leftover roots. It is crucial to understand the breed you are working with to ensure proper treatment.

Rabbits are naturally clean animals, so washing the wool after harvesting is usually unnecessary. It is essential to note that harvesting Angora wool should never cause pain or harm to the rabbit. Any mistreatment is unacceptable and not representative of responsible breeders. Establishing a good relationship with your rabbit, based on trust and familiarity, is crucial before attempting to pluck or shear its wool. This helps minimize the risk of accidental cuts and, in the case of plucking, triggers the release of a hormone that facilitates easier fiber removal.

Benefits of Raising Angora Rabbits

1. They are a no-kill livestock as wool producers, which appeals to many potential rabbit farmers.
2. They do not require a large acreage, making them ideal for urban and suburban homesteaders.
3. Rabbits are cost-effective to feed, consuming only about 4-8 ounces of pellets daily.
4. Breeding is straightforward and reproduction is rapid.
5. Harvesting the wool is a relaxing and enjoyable task that is also essential for the rabbit's well-being.
6. Angora rabbits can be shown in competitions, used as 4H projects, and make wonderful pets, making them a great family activity.
7. The fiber can be sold for profit or kept for hand-spinning by the breeder.

Raising Angora Rabbits

1. Raising Angora rabbits involves specific care practices crucial for their health and fiber production.
2. Consider potential family allergies to animals before deciding to raise them.

3. Regular cleaning of cages is essential.
4. Coat maintenance varies depending on the breed.
5. Daily care, like with any other animal, is necessary.

B. Rabbit Breed For Meat Skin Production

Selecting the right breed of rabbit is essential for successful broiler rabbit farming. With around 89 internationally recognized rabbit breeds, farmers have a wide selection to choose from. In India, breeds like Soviet Chinchilla, Newzealand White, Grey Giant, White Giant, Flemish Giant, Californian, and Silver Fox have shown success in rabbit farming. Each breed has unique qualities that make them suitable for commercial rabbit production. When selecting a breed for broiler rabbit farming in India, factors like climate suitability, resource availability, market demand, and personal preferences should be considered. Farmers should research each breed's characteristics to determine the best fit for their needs and goals. By choosing the most suitable breed(s), farmers can ensure optimal production and profitability in their rabbit farming operations.

1. Soviet Chinchilla: The Soviet Chinchilla is a breed of rabbit known for its blue grey fur with a white belly. It also has a unique physical feature called the 'raff' or 'dewlap', which is a thick fold of skin around the front of its chest that is particularly prominent when the rabbit is well-fed and in good condition. This breed typically weighs between 3-4.5 kilograms when fully matured, making it suitable for meat production. However, its luxurious fur is also highly sought after in the fur crafts industry, making it a popular choice among breeders and enthusiasts alike. With both practical and decorative uses, the Soviet Chinchilla proves to be a valuable and versatile animal within the world of rabbits.

Fig. 8: Soviet Chinchilla

2. New Zealand White: The New Zealand White is a breed of rabbit that originated in England. It is known for its distinct white fur and albino skin. The absence of melanin pigment gives this breed red eyes, making them easily identifiable. With an average weight of 4.5 to 5 kilograms, these rabbits are mainly used for their meat and fur skin, which are highly sought after products. In fact, the New Zealand White is the most commonly used breed for meat production worldwide. Its popularity can be attributed to its large size, docile temperament, and efficient growth rate. This breed has gained recognition globally for its delicious meat and high-quality fur, making it a valuable asset to the rabbit farming industry.

Fig. 9: New Zealand White

3. Giant: Giant rabbit breeds have a long history that dates back to their use for fur and meat production. However, due to their calm and gentle nature, they quickly gained popularity as pets. To be classified as a giant breed, rabbits must weigh an average of 4.5 kg or more. While most giant breeds reach a maximum weight of 11 kg, there are exceptions such as the Valenciano breed from Spain which can grow to the size of a small sheep. Currently, Spain is attempting to revive this breed as it is in danger of extinction. Thirteen popular giant rabbit breeds from around the world are briefly described below.

Flemish Giant: The Flemish breed Rabbit is a large and docile animal, with a weight of up to 11 kg. These rabbits are popular as pets due to their docile nature and love for attention from their owners. Their size also makes them more sturdy and robust compared to smaller rabbit breeds. Females reach their full size at around one year old, while males take about six months longer. With soft, thick glossy hair and an arched back, these rabbits are known for their patience and calm demeanor. Both males and females have broad heads, with the males having a slightly larger head. The coat colours of Flemish rabbits vary from black, blue, light grey, sandy, steel grey, white to fawn. Their large ears stand upright on their heads, adding to their unique appearance.

Fig. 10: Flemish Giant

Continental Giant: The Continental Giant is a breed of rabbit that can weigh up to 7 kg, with some individuals reaching a whopping 11 kg. Despite their size, they are known for being fun, intelligent and gentle animals that love to play. They are also easy to train due to their intelligence and willingness to please. With soft dense fur that can grow up to four centimeters long, these rabbits come in various colors such as dark steel grey, opal, yellow, chinchilla, light steel grey and black. There are two agouti varieties - red and chestnut - as well as two white varieties - blue eyed white (BEW) and red-eyed white (REW). Descendants of the Flemish giants, these rabbits share some of the same characteristics such as a long body that can reach up to three feet in length. Their ears stand up straight from their head and make up 25 percent of their body length. However, despite their large size and impressive appearance, they have a relatively short lifespan of 4-7 years. These gentle giants not only make great human companions but also tend to get along well with other animals. In fact, they may even become best friends with the family cat!

Fig. 11: Continental Giant

Giant French Lop: The Giant French Lop is a notable rabbit breed that can weigh up to 7 kg. This gentle and sociable breed was developed by crossing an English Lop with a French Butterfly Rabbit Breed, resulting in a soft dense coat available in various colors such as blue, brown, and white. With their signature long floppy ears measuring 5 to 8 inches in length, these rabbits have a distinct appearance characterized by a large head, wide forehead, and plump cheeks. Initially bred for meat purposes, the popularity of the Giant French Lop as a pet and companion animal grew significantly in the 1960s. The breed eventually made its way to America in the early 1970s and has since become a beloved addition to many households.

Fig. 12: Giant French Lop

Spanish Giant: The Spanish Giant is a remarkable rabbit breed known for its impressive weight of up to 7 kg. This breed was created by crossing the Lebrel Espanol and Belier rabbit breeds. Originally bred for their meat, they have short and soft fur that comes in various colors. Their body structure resembles that of the Continental Giant, with long ears that stand upright. Despite their large size, Spanish Giants are calm, docile, and friendly animals, making them ideal companions for the elderly. They also get along well with supervised young children and make great family pets. These rabbits thrive on company and must have a furry companion to keep them happy. Due to their size, their lifespan is around 4 to 6 years. Overall, the Spanish Giant is an exceptional breed that not only has an impressive appearance but also makes for a wonderful pet.

Fig. 13: Spanish Giant

Blanc De Bouscat: Blanc De Bouscat, also known as Ermine rabbits, is a large breed of rabbits that can weigh up to 7 kg. This breed was originally developed by Mr and Mrs Dulon in the village of Bouscat in Gironde, France in 1906. It was later renamed Blanc De Bouscat to represent its place of origin. The Dulons achieved this breed by crossing the Flemish Giant with French Angora and Champagne d'Argent, with the aim of creating a pure white rabbit with red eyes. While their primary purpose was for meat and fur production, they have gained popularity as pets over the years due to their calm, affectionate, and playful nature. These intelligent rabbits are now one of the most sought-after pet breeds in France.

Fig. 14: Blanc De Bouscat

Hungarian Giant: The Hungarian Giant rabbit, also known as the Hungarian Agouti rabbit, is a large breed of rabbits that can weigh up to 7kg. These rabbits are characterized by their soft dense fur and were primarily bred for their meat. They were created by crossbreeding various continental rabbit breeds with a few wild rabbit breeds over two hundred years ago. Originally named after their fur color and pattern, the modern Hungarian rabbits now come in a variety of colors and patterns, causing the "Agouti" to be dropped from their breed name. While they are still used for their meat today, they have also gained popularity as show rabbits and pets due to their unique appearance and gentle nature. With its impressive size and striking fur, the Hungarian Giant rabbit continues to capture the hearts of many rabbit enthusiasts around the world.

Fig. 15: Hungarian Giant

British Giant: The British Giant is a highly desirable rabbit breed, known for their impressive weight of up to 7 kg. Not only do they have a hefty size, but they also possess a friendly and easy-going nature. These lovable creatures are often described as inquisitive and even playful, making them a joy to interact with. They thrive on social interaction and can easily bond with humans and other pets, making it essential for them to have a companion. With their long semi-arch Mandolin body type, upright ears, and small cottontails, British Giants have an unmistakable appearance. Their large wedge-shaped heads add to their grandeur. Their fur is soft, thick, and dense, coming in various color varieties such as opal, sable, white, blue, black and steel grey. Due to their love for attention and gentle demeanor, they make excellent companions for the elderly or as house pets for families. Although popular in the UK, this breed may not be widely found in other parts of the world.

Fig. 16: British Giant

Giant Chinchilla: The Giant Chinchilla Rabbit is a large and gentle creature, weighing up to 7 kg. They are appropriately named for their chinchilla color soft dense fur that gives them a striking appearance. These rabbits have a calm and placid personality and enjoy receiving attention from their humans. However, they are not very active and tend to prefer lounging around rather than jumping about. This makes them perfect companion animals for senior citizens, as well as gentle pets for young children. The Giant Chinchilla Rabbit has an elegant blue-grey coat with a white underbelly and a small black tail. Despite their size, these rabbits can live up to ten years, making them long-term companions. They also have a strong sense of loyalty towards their human families and can become possessive of them if allowed to freely roam around. With their large bodies, wedge-shaped heads, and alert upright ears, the Giant Chinchilla Rabbit is certainly an adorable addition to any household.

Fig. 17: Giant Chinchilla

Altex Rabbit: The Altex Rabbit is a unique American breed known for its size, weighing up to 6 kg. Its name is derived from its place of origin, Alabama and Texas, where it was developed at the renowned universities of Alabama A&M and Texas A&M. This breed was created by crossing Champagne d'Argent, Flemish Giant, California, and New Zealand White rabbits. The Altex Rabbit has a distinctive coat with white and dark markings similar to that of Californian rabbits. It has a small head with long straight ears, dark feet, legs, and tail. Although these rabbits are alert and may seem shy at first, they are comfortable with handling. Originally bred for their meat in 1994, the Altex Rabbit is not recognized by the American Rabbit Breeders Association (ARBA).

Fig. 18: Altex Rabbit

Checkered Giant: The Checkered Giant is a strikingly beautiful rabbit with a weight of up to 5.5 kg. Its name is derived from its distinct black and white checkered fur, which is soft, dense and short. The rabbit has a long mandolin-shaped body, a large wedge head, and tall straight ears that are typically entirely black. Despite their size, Checkered Giants have a gentle nature and are known for being friendly, curious, and lovable. They are also highly active, making them great house pets for families as well as companions for the elderly. However, due to their bigger size, they tend to have a shorter lifespan of around 5 to 6 years. Overall, the Checkered Giant makes an excellent pet choice for those who appreciate both beauty and companionship in a furry friend.

Fig. 19: Checkered Giant

Silver Fox: The Silver Fox is a unique and rare breed of rabbit that originated in America through the efforts of Walter Garland from North Canton, Ohio. It was created by crossing different breeds such as Checkered Giants, Champagne D'Argent, and American Blue Rabbits, with a possibility of some English Silvers being involved as well. These rabbits are named after the silver fox due to their strikingly similar fur. Bred for both their meat and their soft, glossy plush fur, they have become a valuable asset to the farming industry. However, their numbers have significantly decreased over the years and they are now listed as threatened by the American Livestock Breeds Conservancy. Currently, they are considered one of the rarest breeds in America with an average weight of up to 5.5 kg.

Fig. 20: Silver Fox

Giant Papillon: The Giant Papillon is a unique and charming breed of rabbit, known for its impressive weight of up to 5.5 kg. One of the most striking features of this breed is their distinctive markings on their soft, dense coat. The white fur is adorned with a black line running down their long spine, giving them a mandolin-shaped body. Additionally, black patches are splotched over their torso, adding to their overall appearance. While they are primarily used as show rabbits due to their beautiful appearance, they also make fantastic pets due to their gentle and friendly nature. However, it should be noted that some Giant Papillons can be temperamental and may become aggressive if not handled properly or in a way they do not approve of. Overall, the Giant Papillon is an exquisite breed with a unique blend of characteristics that make them stand out among other rabbit breeds.

Fig. 21: Giant Papillon

Gray Giant: The gray giant breed is renowned for its massive size and is primarily bred for its high-quality meat and skin. These rabbits typically weigh between 5-7 kilograms and are resistant to diseases, making them a popular choice among farmers in the former USSR countries, as well as in Russia, Ukraine, and Moldova. Efforts are underway to enhance the quality of their coat to further improve the breed. The gray giant breed was developed near Poltava post-World War II by crossing local rabbits with albino Flanders. Through meticulous selection, individuals that met the desired breed standard were chosen. These rabbits are known for their powerful physique, strong bones, oval body with a rounded back, and large fleshy ears set in a V-shape when viewed from the front. While these rabbits typically weigh 5-6 kg on average, record-breaking adult males can reach up to 7 kg in weight.

Fig. 22: Gray Giant

C. Rabbit Breed For Fancy/Hobby Purposes

Rabbits are commonly known for their use in meat and fur production, but there are also breeds that are considered "fancy" and are primarily raised as family pets. These breeds often have unique and desirable physical characteristics, such as long fur or distinct coloring, which have made them popular among rabbit enthusiasts. Unlike breeds raised for meat or fur, fancy breeds are not bred for specific traits but rather for companionship and aesthetics. This makes them excellent choices for families looking to add a furry friend to their home. They have friendly and docile personalities, making them great pets for children and adults alike. Fancy rabbits require relatively low maintenance compared to other pets, making them suitable for busy lifestyles. Additionally, they can be litter-trained and easily kept indoors, providing a clean living environment for both the rabbit and its owners. With proper care and attention, these fancy rabbit breeds can bring joy and companionship to any household.

When considering getting a rabbit for a child, it's best to wait until they are at least 10 years old to ensure they understand the responsibility of caring for a pet. Adult family members should take on most of the care duties. Rabbits are social animals that form strong bonds with their owners, so committing to their care for their 8-12 year lifespan is crucial. Potty training a rabbit is possible with patience, as they have a natural inclination to choose specific spots for urinating and defecating. Providing them with freedom to roam and a litter box can aid in training. Giving rabbits 7-8 hours of cage-free time daily is essential for their physical and mental well-being. They need exercise, mental stimulation, and a variety of toys to keep them entertained. International Rabbit Day on September 26th is a time to celebrate rabbits and raise awareness about their proper care. With patience and the right training, rabbits can be successfully litter-box trained. Providing space to roam and mental stimulation through play and toys are crucial for their happiness and health. Take the opportunity on International Rabbit Day to show your bunny some extra love and care.

Popular fancy rabbit breeds include Polish, Palomino, Havana, Beveren, New Zealand Red, English Spot, and Dutch. These breeds are lighter in weight and known for their fancy appearance.

1. Polish: The Polish rabbit is a compact domestic breed often bred by fanciers and exhibited in shows. Despite its name, it likely originated in England, not Poland. In the UK, it is known as Britannia Petite, while in the US, it is called Polish. The Polish rabbit is popular for exhibition and as a pet due to its small size, short ears that touch from base to tip, short head, full cheeks, and bold eyes. It is slightly larger than the Netherland Dwarf and has distinct differences in coat, body type, and colors. The accepted weight for a 6-month-old Polish rabbit in the US is 1 to 1.5 kg, with the ideal weight being 1 kg. Historically, Polish rabbits in the US were predominantly white with red or blue eyes, with the ruby-eyed white being a true albino and the blue-eyed white having the Vienna gene. Colored varieties were recognized from the 1950s onwards, with black and chocolate colors approved in 1957, blue in 1982, and broken in 1998.

Fig. 23: Polish

2. Palomino: The Palomino rabbit is a breed that typically weighs between 3.6 and 5.4 kg and is known for its golden or lynx coloring. It was first developed in the United States by Mark Youngs at Lone Pine Rabbitry in Coulee Dam, Washington. The breed originated from Youngs' American Beige breed and was initially called Washingtonian in 1952 at the American Rabbit Breeders Association's national convention. The name was later changed to Palomino in 1953, and the breed was officially recognized in 1957, gaining popularity in the 1960s. Palomino rabbits have a fawny brown color, brown eyes, and a larger body size compared to other breeds. They resemble the New Zealand rabbit and come in two main color variations: golden and lynx.

Fig. 24: Palomino

3. Havana: The Havana is a unique and distinct breed of rabbit that originated in the Netherlands in 1898. It has since been bred to create other recognized breeds such as the Fee de Marbourg, Perlefee, and Gris Perle de Hal. The American Rabbit Breeders Association has acknowledged five color variations for Havanas including chocolate, lilac, black, blue, and broken. These rabbits have an average weight ranging from 2.0 kg to 2.9 kg, making them a medium-sized breed. Their smooth and silky fur comes in various shades of brown, making them a popular choice among rabbit enthusiasts. With their long history and recognized breed status, Havanas are a treasured addition to any rabbitry or pet owner's collection.

Fig. 25: Havana

4. Beveren: The Beveren rabbit, originating from the small town of Beveren near Antwerp in Belgium, is one of the oldest and largest breeds of fur rabbits. They come in various colors such as blue, white, black, brown, and lilac, but only three are recognized by the American Rabbit Breeders Association (ARBA). One rare variety known as Pointed Beveren has white-tipped hairs in these same colors. The blue variety is considered to be the original. These rabbits are known for their calm temperament, hygiene habits, and intelligence. With a strong desire for outdoor exploration, they are full of energy and make excellent pets. Their dense and glossy coat should have a gentle rollback fur type. This large breed has a distinct mandolin shape and can weigh between 3.6-5.4 kg pounds for senior bucks and does. Their offspring usually have large litters that grow quickly under the care of docile does.

Fig. 26: Beveren

5. New Zealand Red: The New Zealand Red rabbit breed, despite its name, actually originates from the United States. Initially developed in California in 1910, these rabbits were originally bred for their fur and meat but are now cherished as pets. The exact lineage of the New Zealand Red rabbit is not fully known, but it is believed that the now-extinct Golden Fawn breed, a hybrid of Flemish Giant and Belgian Hare, played a significant role in its creation. It is said that Golden Fawns were crossed with Belgian Hares to achieve the striking ginger fur of the New Zealand Reds. While the British Rabbit Society recognizes the New Zealand Red as a distinct breed, the ARBA considers it a color variation of the New Zealand Rabbit breed. Known for their vibrant ginger coat and friendly demeanor, New Zealand Red rabbits make excellent family pets. Weighing between 3 to 4.5 kg, New Zealand Red rabbits are considered a large breed. They have a sturdy, semi-arched body shape with long, upright ears and a rounded head. Their short, flyback fur requires minimal grooming, with a weekly brushing being sufficient to keep them tidy. During molting season, they may shed more and require more frequent grooming. The New Zealand Red rabbits have a deep, vibrant cinnamon-red coat color with no patterns or markings, complemented by creamy beige bellies and brown eyes. These rabbits are affectionate, social, and enjoy being handled by their owners.

Fig. 27: New Zealand Red

6. English Spot: The English Spot is a medium-sized domestic rabbit breed that originated in England in the 19th century through selective breeding. They typically weigh between 2.2 to 3.6 kg and are known for their unique colored markings, including butterfly nose marking, eye circles, cheek spots, herringbone, colored ears, and a chain of spots. The breed has a flyback fur type and comes in seven varieties: black, blue, chocolate, lilac, tortoise, gray, and gold. English Spots have a distinctive full arch body with long front legs that lift them off the table, and they are recognized for their curious and playful nature. The breed has been popular in England since the 1850s and was introduced to North America around 1910. The American English Spot

Rabbit Club was established in 1924, and the French named the breed "Lapin Papillon Anglais" or the English Butterfly Rabbit due to the butterfly marking on the nose. Over the years, the breed has developed distinct and well-defined markings.

Fig. 28: English Spot

7. Dutch: The Dutch rabbit is easily identifiable by its white markings, which include a blaze on the nose, a collar, and a saddle on the back. They are a small breed, weighing 3.5 to 5.5 pounds, and have a lifespan of 8 to 10 years. Their fur is of normal length, with a soft underlayer and longer guard hairs, and their ears are upright. The breed standard describes a compact, rounded, and balanced body with distinct markings. The American Rabbit Breeders Association recognizes seven color varieties of Dutch rabbits, all with white markings: black, blue, chinchilla, chocolate, gray, steel, and tortoise, with a lilac color in development. Eye colors can range from brown to blue-gray.

Fig. 29: Dutch

The Dutch breed has a long history, originating from the Brabancon breed in Belgium and acquiring its name by 1835. Dutch rabbits are known for their easygoing, friendly, and intelligent nature, though individual personalities may vary. To build trust with a Dutch rabbit, observe their behavior and allow them to approach you on their terms.

4

Housing and Management Practices

The success of a rabbitry depends heavily on the equipment and hutches used. Climate, location, and budget all play a role in determining the type of equipment needed. Before constructing the hutches, careful planning is necessary to minimize labor and ensure optimum working conditions. A clean and efficient layout should also be considered in order to provide an ideal environment for raising rabbits. With a wide range of hutch designs available, it is important to choose one that best suits the specific needs and purpose of the rabbitry. There is no one perfect design that can cater to every situation or function, so proper consideration must be given before selecting the appropriate equipment and hutch for a successful rabbitry operation.

In addition to offering the rabbits optimal comfort through sufficient shelter from all types of weather and predators, hutches should have good ventilation. Adequate ventilation is essential for the health and well-being of animals, as it helps to maintain a clean and dry living environment. This protects the rabbits from various weather conditions such as strong winds, typhoons, and excessive sunlight exposure. When setting up a rabbit farm, it is important to consider using local resources whenever possible to reduce costs and support sustainable production. It is also crucial to integrate rabbit farming into regional livelihood systems to maximize its benefits for the community. By utilizing materials already available in households, such as bamboo and scrap lumber, farmers can create durable and cost-effective hutches for their rabbits. Proper shelter with good ventilation will not only ensure the well-being of the animals but also contribute to successful rabbit production.

Wire Hutches

Wire hutches are a popular choice for many rabbit owners due to their versatility and durability. These hutches often incorporate different sizes of chicken wire, although the wear and tear on this material is high. For added safety, rabbit hutches may have their walls or ceilings covered with a thin gauge of chicken wire, preventing small kits from escaping the nest box and staying within the hutch's enclosed space. It is recommended to use a diameter less than an American quarter to ensure the rabbits' safety. However, it is not advised to cover the floor with chicken wire as it can cause painful hocks and other sores on the rabbits' feet, leading to infections and overall poor output. In terms of size, a minimum dimension of 30 by 30 by 18 inches is suitable for cages: 30 inches wide, 30 inches deep, and 18 inches high. This ensures enough space for the rabbits to move around comfortably while also promoting good health and well-being.

The construction of a hutch for rabbits should be carefully planned and executed, with consideration given to both the size and materials used. The sides and top of the hutch should be covered with one-by-two-inch 14-gauge galvanised wire, while the floor should be covered with half-by-one-inch 14-gauge galvanised wire. It is preferable that the wire is galvanised after weld for better durability. Depending on the available space, the hutch can be set up in single, double, or triple tiers. A waist-high, single-tier hutch is ideal if there is enough space as it allows for easy observation of rabbits and requires less time and effort for feeding and care. However, a combined two-compartment all-wire hutch may cost more to construct despite being easier to build. A three-tier configuration requires more work and time as caring for rabbits at different levels can be challenging. This design also compromises proper sanitation. The recommended schematic includes three hutches with "doors" carved out of each unit for easy access to all compartments in one go. Overall, careful planning and consideration are crucial in constructing an efficient hutch for rabbits.

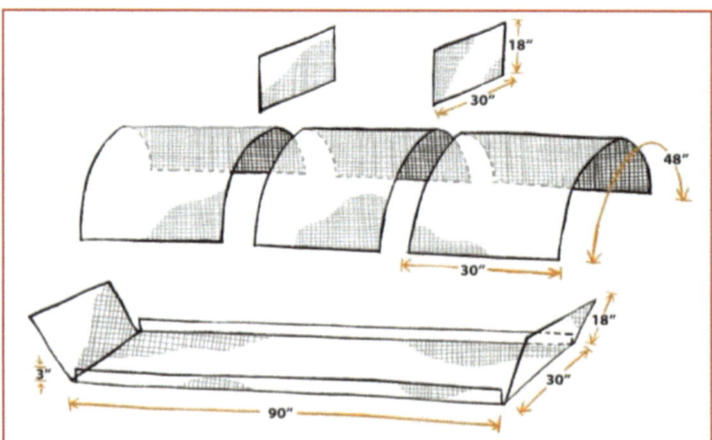

Fig. 1: Three-hutch System

A hutch in the shape of a Quonset, constructed entirely from wire, offers numerous advantages over traditional rectangular hutches. Not only does it require less material, thus saving costs and promoting environmental sustainability, but its distinctive structure also makes cleaning a breeze. Unlike typical hutches with tight corners and hard-to-reach areas, the Quonset-shaped hutch eliminates these challenges by providing a smooth and open surface that is easy to access and maintain. Additionally, the top door feature allows for convenient feeding and cleaning without having to navigate through a small doorway. This design also eliminates the need to contort oneself to reach into the corners of a single-tiered hutch at waist height. This makes maintaining hygiene and caring for rabbits more comfortable and less physically demanding.

Moreover, the adaptability of these hutches makes them an excellent choice for various rabbitry setups. They can be customized to suit different needs and preferences, making them versatile and adaptable solutions for housing rabbits. Constructing two-hutched units using this design is a simple process that requires minimal effort. Overall, a Quonset-shaped hutch made entirely from wire not only offers practical benefits but also promotes ease of use and efficiency in rabbit care.

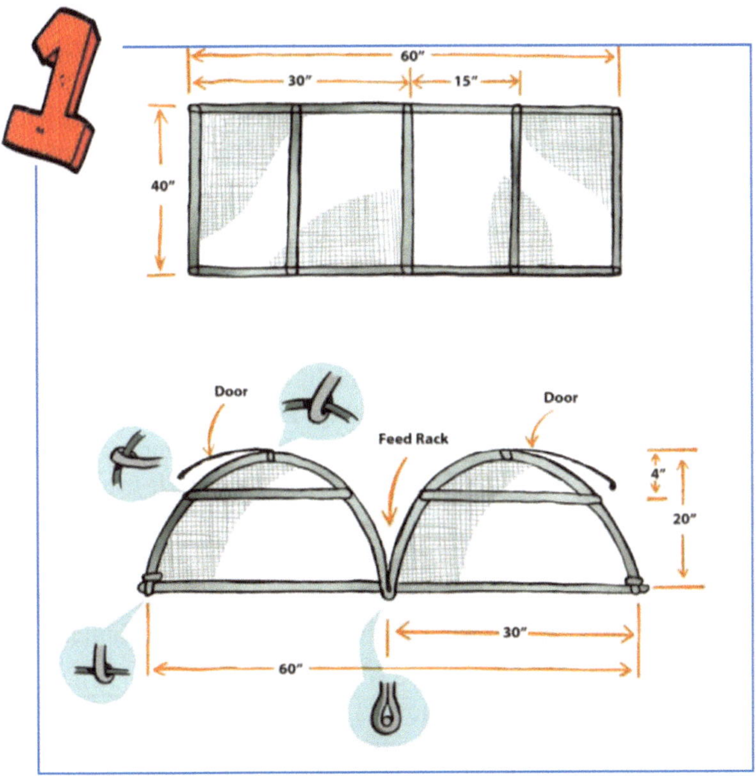

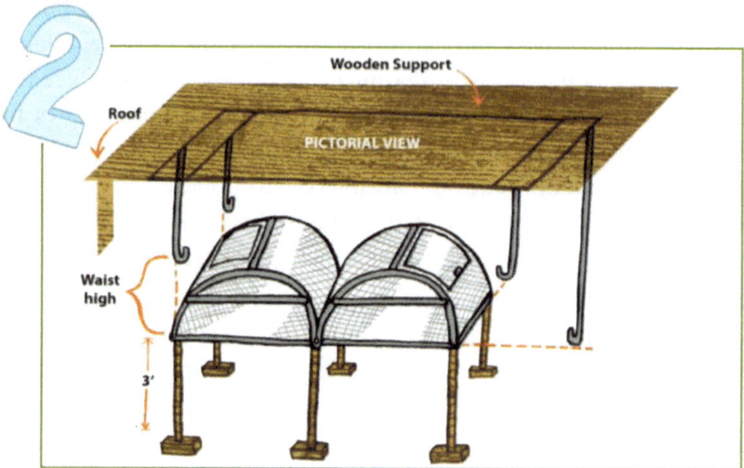

Fig. 2: Two-cage, All Wire Quonset Hutch

Wood-frame and Wire Hutches

Wood-frame and wire hutches are a popular and cost-effective option for housing rabbits. These types of hutches consist of a wooden frame and welded wire sides, floor, and top. The wooden frame serves as the external skeleton, while the welded wire provides superior ventilation and maintains exceptional sanitation.

One of the key advantages of this type of hutch is its longevity. The use of durable materials such as wood and wire ensures that the hutch will last for many years with proper maintenance. Additionally, the open design allows for ample air circulation, promoting good health in rabbits.

When constructing a wood-frame and wire hutch, it is important to consider hygiene. To make cleaning easier, ensure that the corner posts are long enough to allow access underneath the hutch. Placing cement blocks under the corner posts can also help extend the life of the wood by preventing it from coming into contact with damp ground.

In areas where ants may be a concern, additional measures can be taken to protect the hutch. Hollowing out sections in cement blocks and filling them with insect repellent can deter ants from crawling onto or into the hutch.

There are multiple options for supporting a wood-frame and wire hutch within a shed or building. It can be hung from rafters or ceilings using heavy wire or light lumber, or rested on a crosspiece between studs. Careful consideration should be given to ensure that the support structure is strong enough to hold the weight of both the hutch and its inhabitants.

Overall, wood-frame and wire hutches offer an affordable yet durable housing option for rabbits. With proper construction and maintenance, these types of hutches can provide a safe and comfortable living space for your furry friends.

Fig. 3: A wooden-framed hutch with steel wire mesh

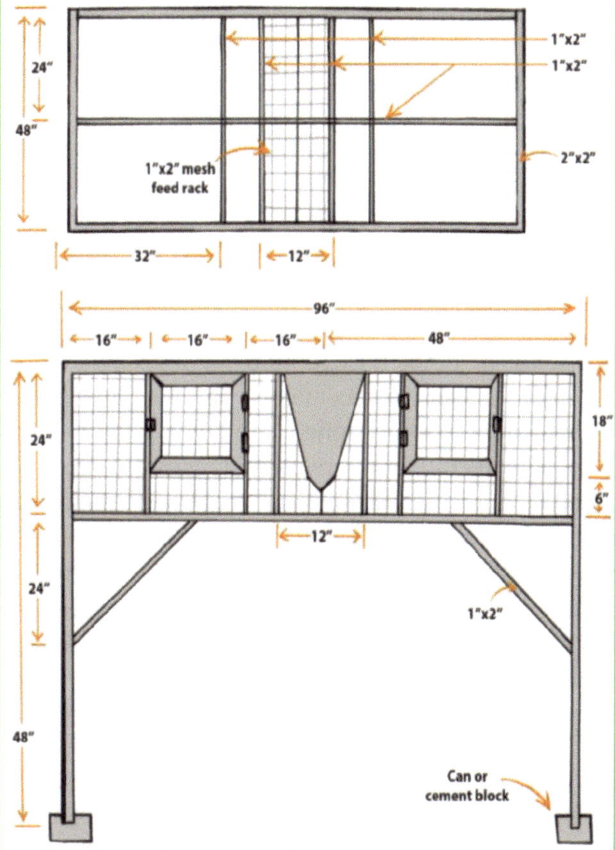

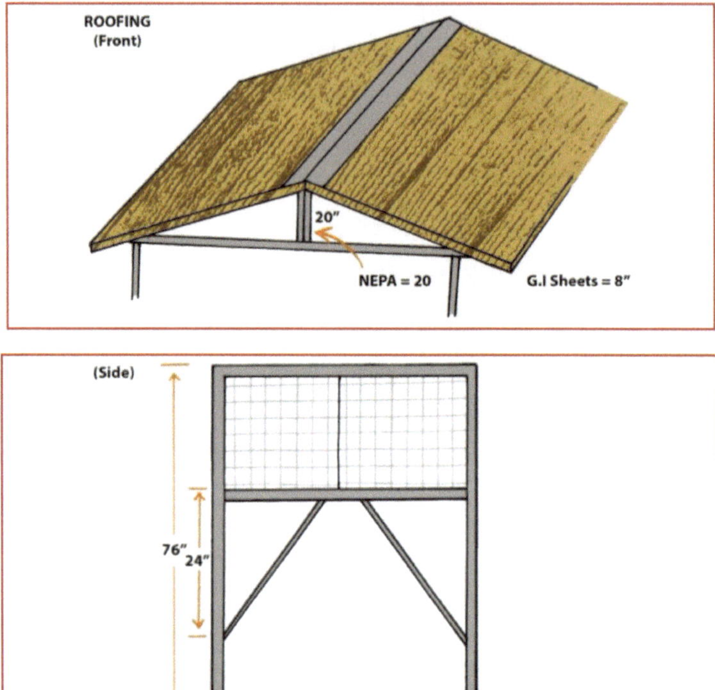

Fig. 4: Double-Faced, Wood-Framed Hutch with Galvanized Wire

Bamboo Hutch

The Bamboo Hutch is a highly recommended option for housing rabbits, offering both cost-effectiveness and environmental friendliness. Its unique design features a roof made of nipa or cogon grass, providing a natural and sustainable source of shelter. The overall structure closely resembles that of a traditional wire-framed wooden hutch, with the added benefit of bamboo slats spaced half an inch apart for the flooring and sides. It is crucial to position the rounded side of the bamboo slats towards the inside of the cage to prevent rabbits from chewing on them. In addition, using straight bamboo for the flooring prevents potential injuries to your rabbit's feet. With its sturdy and organic materials, the Bamboo Hutch creates a comfortable and secure living space for your beloved furry companion.

Hutch Floor

Hutches are commonly used for housing rabbits, and their flooring plays a crucial role in maintaining the cleanliness and comfort of these furry companions. There are various types of flooring options available for hutches, each with its own unique benefits.

One popular choice is wire-mesh flooring, which is ideal for self-cleaning purposes. This type of flooring is especially useful in commercial settings with limited labor, as it allows for easy maintenance and promotes good hygiene. However, when installing wire-mesh flooring, it is important to thoroughly check for any sharp points that may result from the galvanizing process. These sharp points can potentially harm the rabbits and should be covered with iron paint to prevent corrosion.

Another option for hutch flooring is solid floors. While easy to maintain at first, they can become unsanitary over time if not properly cared for. To improve their hygiene levels, it is recommended to slope them slightly for proper drainage. This will prevent any stagnant water or waste from accumulating on the surface.

Bamboo or oak slats can also serve as suitable choices for hutch flooring. These materials are durable and offer a natural aesthetic to the hutch. For a balanced approach, some owners choose to combine a solid front floor with wire-mesh or slats at the back. This allows for both comfort and easy cleaning.

In addition to considering hygiene and maintenance, it's important to also prioritize the rabbit's comfort when choosing hutch flooring. One way to ensure this is by adding a plywood resting board sized appropriately for the animal's needs. This will provide a comfortable surface for them to rest on while preventing any potential foot sores.

Overall, investing in appropriate hutch flooring is essential for creating a safe and comfortable environment for your rabbits. With proper care and consideration of different options available, you can provide your furry friends with an ideal living space they will enjoy spending time in.

Feeding Equipments

Feeding equipment is a crucial aspect of caring for rabbits, and it is important to encourage creativity when considering options. This is especially true when cost is a concern, as utilizing locally accessible resources can be a more affordable solution. In order to assist in this innovation process, here are some key points to keep in mind:

1. Take into consideration food leakage and waste that may occur with small kits crawling into low open top options like troughs.
2. Be mindful of food scraps left on the hutch's floor, as they can quickly become dirty and potentially harmful to your rabbits.

3. It is also worth noting that certain rabbits may develop the habit of chewing feeder items, particularly plastics or wood/bamboo. However, it should not be assumed that all rabbits will exhibit this behavior.
4. To save time and efficiently feed multiple rabbits, it is best to choose feeders that are large enough to accommodate multiple feedings.
5. When selecting a feeder, make sure it will prevent waste and contamination from occurring.

Remember to approach the feeding equipment selection process with care and attention to detail. By taking these points into consideration, you can ensure that your rabbits receive proper nutrition without any unnecessary waste or risk of contamination.

Types of Feeders

1. Crocks

Rabbit-feeding crocks are a convenient and practical option for providing rabbits with their daily intake of food. Measuring approximately 6 inches in width and 3 inches in height, these crocks are compact in size and ideal for small spaces. With a concave lip design, they effectively prevent rabbits from wasting their feed by clawing it out. Additionally, their sturdy construction ensures stability to avoid any tipping over. Despite the numerous benefits, there have been concerns raised about the potential contamination of feed when baby bunnies enter the crock. To address this issue, it is recommended to choose materials such as pottery or earthenware for the crocks. These materials are known for their durability and non-toxic properties, making them safe for the health of rabbits.

Furthermore, rabbit-feeding crocks offer convenience to pet owners by reducing mess and waste compared to traditional feeding methods. They also promote better portion control, ensuring that rabbits receive just the right amount of food per serving. When using these crocks as a rabbit feeder, it is important to regularly clean them to maintain proper hygiene standards. This will prevent any bacteria or mold growth that could potentially harm your furry friends. In addition to providing food, rabbit-feeding crocks can also be used as a water source for rabbits by filling them with clean drinking water. Some models even come with hooks or clips for easy attachment to cages or enclosures.

Whether you are a new rabbit owner or an experienced one looking for a more efficient feeding method, consider using specialized rabbit-feeding crocks. Their compact size, waste-preventing design, and multiple uses make them an excellent choice for keeping your bunnies well-fed and healthy. Remember to choose durable materials and maintain regular cleaning habits to ensure optimal feeding conditions for your furry companions.

2. Grass Mangers

Grass mangers are essential structures for rabbit care and are typically constructed using 1-by-2-inch mesh wire with a gauge of 16. These mangers can be shaped in either a U or V form to provide rabbits with easy access to the grass by pulling it through the mesh. Alternatively, they can be built at the front or side of the cage, though this may require more effort from both the animal and its caretaker. It is recommended to position grass mangers between two cages as this not only saves space but also reduces labor. This placement allows multiple rabbits to access the manger simultaneously, promoting efficient feeding practices. The 1-inch vertical wire and 2-inch horizontal wire arrangement enables rabbits to easily pull and consume grass from the mangers. When constructing grass mangers, it is important to use sturdy materials such as high-quality mesh wire with a gauge of 16. This ensures durability and prevents rabbits from escaping or damaging the structure. Regular maintenance checks should also be conducted to ensure that the mangers remain intact and safe for the animals.

Overall, grass mangers play a crucial role in providing rabbits with easy access to fresh grass, promoting their well-being and health. By following proper construction methods and regularly monitoring their condition, these structures can effectively contribute to successful rabbit husbandry practices.

3. Bamboo Troughs

Bamboo troughs are a useful and sustainable option for providing feed to rabbits in a cage setting. By following the steps below, you can easily create a functional and secure rabbit feeder using bamboo troughs.

1. Begin by selecting two nodes on your bamboo trough that are closest to each other in terms of size and shape.
2. Using a sharp knife or saw, carefully cut away one-third of the side of the bamboo between these two nodes. This will create a concave container that is ideal for holding rabbit food.
3. Ensure that all edges and corners are smooth to prevent any potential injury to the rabbits while they eat.
4. Next, securely attach the feeder to the bottom or side of the cage using wire. This will prevent it from toppling over and causing any spills or messes.

5. Fill the feeder with appropriate rabbit food, taking care not to overfill it.
6. Monitor your rabbits' eating habits and adjust the amount of food in the feeder accordingly.
7. Regularly clean and refill the feeder to ensure your rabbits have access to fresh feed at all times.

Using bamboo troughs as rabbit feeders not only benefits our furry friends but also promotes environmentally friendly practices through sustainable use of natural resources. Give this method a try for an efficient way to provide feed for your rabbits while keeping their cage organized and tidy.

4. Hoppers

Raising rabbits for their meat, fur, or as pets requires efficient and effective feeding practices. One recommended method is the use of hoppers as rabbit feeders. These specially designed containers save valuable time and labor while providing a convenient and tidy way to store and dispense feed. Hoppers can be constructed using readily available materials such as ceramics, metal, or wood. They are typically placed inside the rabbit hutch or suspended outside of it, keeping the feed dry and accessible for the rabbits. One important consideration when designing a hopper is its size. It should be able to hold enough feed to last several days, reducing the need for frequent refilling. Additionally, for ease of consumption by young bunnies, the entrance where they receive their feed should not be more than 4 inches from the floor of the hutch.

In terms of feed type, complete rabbit pellets are highly recommended for use in hoppers. These provide a balanced diet for rabbits and can easily be dispensed through the small openings in the hopper design. For those looking for a cost-effective option, a simple hopper can easily be constructed using materials such as plywood or lawanit (fibre board). Alternatively, a standard square 5-gallon container can also serve as an effective hopper and has the capacity to store approximately 15 pounds of pellets or home-mixed feeds.

By utilizing hoppers as rabbit feeders, you can streamline your feeding process and ensure that your rabbits receive proper nutrition without wasting any excess food. This translates to healthier rabbits and more efficient management of your rabbitry operations.

To properly construct a hopper for feeding small animals, follow these refined steps:

1. Begin by cutting off the top of a five-gallon can using appropriate tools.

2. On two opposing sides of the can, make incisions to create openings for the feed to flow through. If you plan on hanging the hopper, also make a hole on one side near the top for attaching it to the hutch. Space these holes evenly at one inch apart and make them four inches high from the bottom of the can. After cutting, bend any rough edges inwards to ensure strength and smoothness.
3. Using a 1x4x13.5 inch board, cut two equal triangles for support.
4. Attach 1/4-inch plywood baffle boards below each opening on the can's sides to guide the feed flow. It is important to have a snug fit against the inside of the can and secure the upper corners against its edge.
5. To prevent any nibbling on exposed edges, cover them with tin or similar material.
6. Secure one of the support triangles onto the baffle board using finishing nails.
7. Once secured, bend one nail at an angle so that it hooks over and rests on the lower lip of each opening for added stability.
8. For those looking to save space, consider using a hopper with a single feed opening instead. To do this, make a hole in one side of your hutch for attaching the hopper and use wire for additional stability on top. You may also want to add a small baffle inside this type of hopper to prevent feed from accumulating in corners.

Following these steps will result in a well-constructed and effective hopper for your small animal's feeding needs.

When a single type of feed is provided for rabbits, it is necessary to use a one-compartment feed hopper. This ensures that the rabbits only have access to the specific mixed feed that is designated for them. It is important for the rabbits to be able to differentiate their food in order to maintain a balanced and healthy diet. However, there can be consequences if this system is not properly managed. Any uneaten sections of the feed can become wasted and may also be scratched out by the rabbits. Additionally, if moisture enters into the container and comes in contact with the feed, it can lead to mold growth.

Moldy feed poses a significant risk to rabbit health as it can cause a build-up of fluids and gas in their digestive system which they are unable to expel. This can result in bloat and eventual death of the animal. Therefore, it is crucial to regularly check and clean the feed hopper and ensure that no moisture or mold has accumulated. Taking proper precautions during rainy weather is especially important as excess moisture can easily enter the container. This includes ensuring that the hopper is sheltered from rain, covering it with a waterproof material, or moving it indoors during wet conditions.

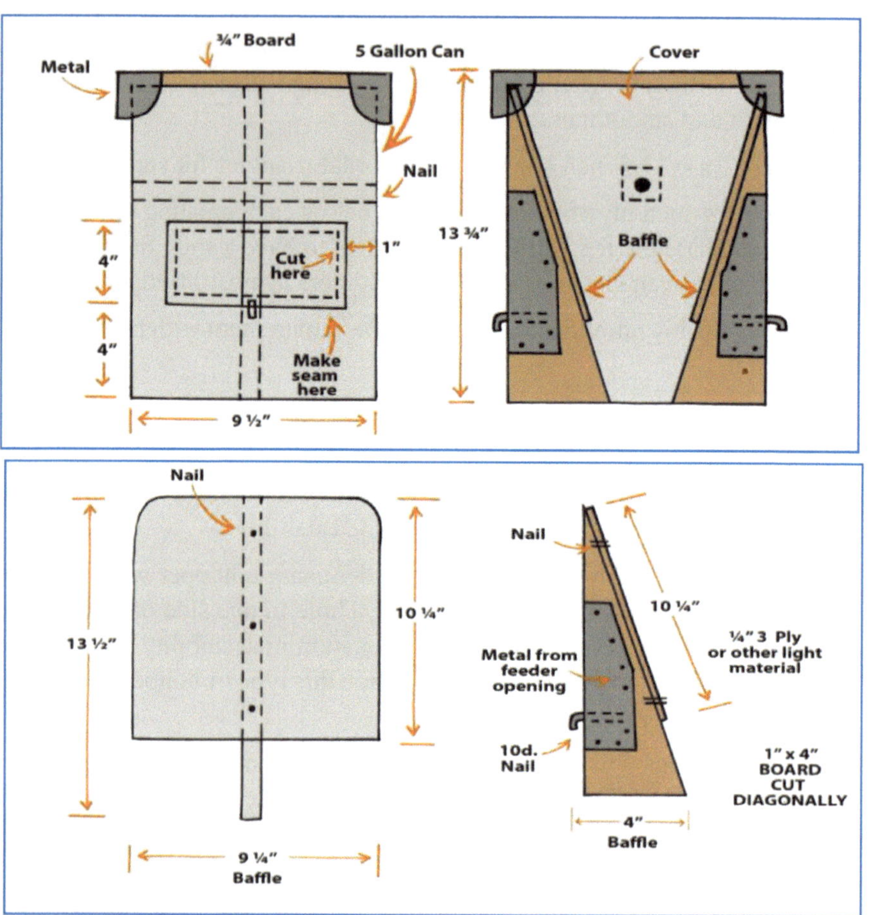

Fig. 5: Feed Hopper Designs Composed of a Five-Gallon Can

It is important to be cautious when using a single-compartment feed hopper for rabbits, as any contamination or spoilage can harm them. Regular monitoring and maintenance of the hopper are essential to ensure the rabbits receive clean and nutritious food, free from mold or other contaminants. A common issue when dealing with litters of bunnies is the spoilage of uneaten feed due to the presence of faeces and urine. Bunnies naturally excrete waste, and when they climb into feed hoppers, their urine can damage the feed. Additionally, the salt found in pelleted rabbit feed can cause metal cans to corrode over time. If mash feed is used instead, the little particles sticking to the sides of the container can also contribute to a higher rate of corrosion. To prevent this costly problem, a cost-effective solution is to paint the metal cans with iron paint. This simple yet practical approach can significantly reduce rusting and prolong the lifespan of feed containers.

Waters

Water is an essential component of a rabbit's daily needs, and it is crucial for their health and well-being. Despite popular beliefs and practices, rabbits should always have access to clean and fresh water at all times. This is especially important during hot weather when their water intake increases significantly.

A single doe, along with her litter of approximately six to eight young rabbits, will consume around two liters of water per day during hot weather. This highlights the importance of providing an adequate amount of water for rabbits, as it directly impacts their daily water requirements. Not only does clean and fresh water help maintain proper hydration levels in rabbits, but it also aids in digestion and helps flush out any toxins from their system. Additionally, rabbits are meticulous groomers who require a sufficient amount of water to keep themselves clean. Providing clean and accessible sources of drinking water also prevents potential health issues such as dehydration, urinary tract problems, and gastrointestinal stasis in rabbits. These conditions can be detrimental to a rabbit's health if left untreated. Furthermore, it is crucial to regularly check the quality of the water provided to ensure that it meets suitable standards for consumption. Water should be changed daily or more frequently if necessary. Any signs of contamination or discoloration should warrant immediate action to prevent potential harm to the rabbits.

Types of Water Containers

a) **Enamel Cups:** Enamel cups are excellent drinking vessels for rabbits because of their superior hygiene and easy cleaning. They offer a more sanitary option compared to traditional crocks. With a quick wash, you can ensure your rabbits have clean water, reducing the risk of health issues. One of the main advantages of using enamel cups is their easy cleaning process. Unlike crocks that may need scrubbing or soaking to remove bacteria or residue, enamel cups can be washed with hot soapy water and dried thoroughly before refilling. This saves time and ensures a continuous supply of clean water for your rabbits. To prevent spills, it's recommended to securely tie the enamel cups to the side of the cage. This keeps the cup in place and minimizes messes inside the cage. Enamel cups are durable and long-lasting, making them a cost-effective choice for rabbit owners. They are resistant to scratches and damage, ensuring they remain in good condition for extended periods.

Fig. 6: A water cup made of enamel that is connected to the cage to stop spills

b) **Bamboo Troughs:** Bamboo troughs are a popular option for providing drinking water to rabbits due to their accessibility. However, they can pose some challenges as algae growth is often a concern.

While bamboo troughs offer a convenient solution for keeping rabbits hydrated, their natural materials make them susceptible to algae growth. Algae can accumulate on the inner walls of the troughs, potentially contaminating the water and making it unappealing to rabbits. To address this issue, regular cleaning of the troughs is necessary. A simple scrub with hot water and mild detergent can remove any algae buildup and maintain the cleanliness of the trough. It is recommended to clean the trough at least once a week or more frequently if needed. Additionally, placing the bamboo trough in a shaded area can help prevent excessive algae growth, as sunlight promotes its development. Regularly changing out the drinking water can also reduce algae growth and ensure that fresh, clean water is always available for your rabbits. In case of heavy algae infestation, using a small amount of vinegar or bleach (carefully rinsed off) can effectively kill off stubborn algae colonies. However, be sure to thoroughly rinse out all traces of cleaning agents before refilling with drinking water.

c) **Crocks:** In the rabbitry, earthenware crocks serve as a practical and cost-effective solution for providing drinking water to rabbits. These vessels, made from fired clay, offer a hygienic option for ensuring the health and well-being of your rabbits. Not only do earthenware crocks come at an affordable price point, but they also have the added benefit of being easy to clean and maintain. This is especially important in a rabbitry setting where cleanliness is crucial for preventing illness and promoting overall health. These crocks can hold an ample amount of water, making them ideal for housing multiple rabbits. Additionally, their sturdy construction ensures they can withstand the daily wear and tear of being used by active rabbits. Furthermore, earthenware crocks are designed to keep the water cool and fresh, which is essential for maintaining proper hydration in your rabbits. Their thick walls provide insulation against external temperature changes, ensuring that the water remains at a suitable temperature for your rabbits to consume. Utilizing earthenware crocks as drinkers for rabbits is not only a practical choice but also a responsible one. Their affordability, ease of maintenance, and ability to keep water cool make them an excellent choice for promoting the health and well-being of your beloved rabbits in the rabbitry.

d) **Ceramic Crocks:** Ceramic crocks are a popular choice for rabbit drinkers due to their durability and ease of maintenance. These containers, made from ceramic materials, provide a safe and sanitary option for rabbits to access fresh water at all times. Ceramic crocks are suitable for use at any stage of a rabbit's life, from young kits to full-grown adults. This is because they are sturdy enough to withstand chewing and other rough handling. They also have a weight that prevents them from being easily tipped over, reducing the risk of spills and contamination. Additionally, these crocks are non-porous which means they do not absorb dirt or bacteria. This makes cleaning them an effortless task - simply wash with soap and water and allows drying before refilling with fresh water. Furthermore, ceramic crocks come in various sizes and shapes, making it easy to choose one that best suits your rabbit's needs. For larger groups of rabbits or outdoor enclosures, larger capacity crocks can be used while smaller sizes may suffice for individual cages. These benefits make ceramic crocks an ideal choice for providing adequate hydration for your rabbits. And if they fall within an affordable price range, they are highly recommended as a long-term investment in your rabbit's health and well-being. When selecting a water dispenser for your rabbits, consider ceramic crocks for their durability, easy maintenance, non-porous surface, and range of sizes. Your rabbits will appreciate the quality and convenience for their health and well-being.

e) **Automatic Waterer:** Automatic watering systems have become a standard practice in commercial rabbitries in the United States. Despite the initial cost, these systems can be adapted for use in rabbitries in disadvantaged nations. Compared to traditional water containers, automatic systems are superior, eliminating the need for manual filling, rinsing, disinfecting, and cleaning. They ensure rabbits have constant access to clean water without the risk of clogging from fur or grime.

While automatic watering systems offer many benefits, there are some drawbacks to consider. Rabbits may need time to adjust to the system, leading to a temporary decrease in water consumption that could affect production. Improper installation and maintenance could also result in leaks that may damage wire mesh.

Despite these challenges, the advantages of automatic watering systems for rabbits outweigh the drawbacks. With proper installation and upkeep, these systems offer a more efficient and reliable way to provide clean water to rabbits in commercial rabbitries worldwide.

f) **Cans:** Cans can be utilized as drinkers for rabbits, however, it is important to note that there are potential risks involved. Rabbits may consume the rust that develops on cans, which can lead to health issues. As a result, it is not recommended to use cans as drinkers for rabbits.

   Alternatively, you can repurpose a 1-liter plastic oil container as a feed or water container for your rabbit. Before using it, make sure to thoroughly clean and sterilize the container to ensure the safety of your pet. Cut an appropriate sized hole in the top of the container for easy access to food or water. To prevent spills and keep the container securely attached to the cage, you can use wire to secure it in place. This will also prevent your rabbit from knocking over the container and making a mess.

Rabbit Housing Options for Optimal Care and Management:

There are many options available when it comes to housing rabbits, all of which have their own benefits and drawbacks. As a responsible rabbit owner, it is essential to carefully assess your needs and the well-being of your rabbits in order to select the most suitable housing system. Professionals in the industry often utilize various housing systems such as hutches, wire-bottom cages, colony setups, and indoor enclosures. Each option offers different advantages such as protection from predators, ease of cleaning, social interaction opportunities, and space for exercise. It is vital to consider factors such as climate, space availability, number of rabbits being housed, and level of care required before making a decision on rabbit housing. Ultimately, selecting an appropriate housing system will contribute towards optimal care and management for your rabbits. Here are some recommended housing options that can be utilized in rabbit husbandry

Cage System

The most efficient way to house rabbits is through a cage system that can either be stored outside on concrete, metal, or wooden racks, or housed within a shed. The shed approach offers a more permanent solution with the use of a concrete floor for proper drainage and a sturdy brick and iron half-wall. However, for those looking for a more cost-effective option, a semi-permanent building can be constructed using locally available materials such as wooden planks and pillars. Inside the shed, cages are arranged on racks, with a common corridor connecting each row. While cages come in various sizes, the standard dimensions are 90 cm by 65 cm wide by 40 cm high or approximately 0.6 square meters. This size provides enough space for rabbits to move around comfortably while also minimizing overcrowding. To optimize space and minimize effort in managing the rabbit housing system, cages can also be stacked in layers inside the shed. This tier system allows for better utilization of vertical space while still providing adequate living space for the rabbits. Overall, this cage system is designed to provide a safe, comfortable and efficient housing solution for rabbits while also considering practicality and cost-effectiveness. Suitable for both indoor and outdoor use, it ensures that rabbits are kept in optimal conditions for their wellbeing and productivity.

Hanging Cage System

The hanging cage system for rabbits is a unique and efficient method of housing these animals. It is similar to traditional cage systems, but instead of being placed on the ground, the cages are suspended from the roof using chains or ropes. This allows for better utilization of space and can be installed in larger structures such as sheds or barns. One advantage of the hanging cage system is that it eliminates the need for a solid flooring. The cages can be hung over natural earth, which reduces the cost and effort of installing a separate flooring. Additionally, with this system, the feces from the rabbits will remain on top while urine can easily be absorbed by the soil below. Not only does this method improve sanitation and waste management, but it also minimizes the risk of disease transmission between rabbits. This is because each rabbit has its own individual hanging cage, reducing any physical contact and potential spread of illnesses like snuffles, rabbit viral hemorrhagic disease, or respiratory issues. Furthermore, due to limited socialization in this type of housing, coccidiosis is not a concern. Rabbits housed in hanging cages do not interact with each other as closely as they would if they were placed on the ground. Therefore, there is a lower risk of contracting this common gastrointestinal disease.

Incorporating hanging cage systems into your rabbit farming operation offers numerous benefits that can greatly improve efficiency and overall success. These innovative systems optimize space utilization, simplify waste management, and reduce potential health hazards associated with communal living. Ensure the success of your rabbitry by implementing this modern approach within your shed or barn.

Fig. 7: Hanging cage system

Hutch System

Another highly effective option for housing rabbits is a hutch system. While it may require a larger financial investment compared to other types of dwellings, hutches offer several benefits that make them well worth the cost. These structures can be constructed from various materials such as wood, bamboo, or iron and should have wire mesh flooring for easy cleaning. It is important to ensure that each compartment within the hutch meets the minimum dimensions required for comfortable living space. One unique benefit of a hutch system is its flexibility. By incorporating movable chambers within the design, owners can easily relocate their rabbits to different areas as needed. When using wood to construct the cages, it is crucial to cover it with mesh wire on the inside as rabbits have a tendency to chew on wood surfaces. During hot summer months, it is recommended to keep hutches in shaded areas to prevent heat stress in rabbits. Unlike other types of housing that may require proximity to a barn or small building, hutches are standalone structures that offer independence and convenience. In order to promote optimal comfort and health for rabbits, wire flooring should be used in hutches. However, this may increase the risk of sore hocks. To avoid this issue, natural bedding or stress pads can be added as an extra layer of protection and cushioning for their feet. Overall, a well-designed hutch system offers an ideal balance between durability, functionality, and comfort for rabbits while also providing convenience and ease of maintenance for owners.

Fig. 8: Hutch system of Housing

Colony Housing System

The use of colony housing has become a popular alternative to conventional cages in rabbit farming. This system allows for rabbits to interact and socialize with one another, promoting a more natural and enriched environment. Each rabbit in a colony requires a minimum floor area of 0.4 square metres, providing ample space for movement and exercise. Within the colony housing system, designated areas are created for does, bucks, and kits. This separation helps prevent fighting between rabbits and any undesired mating. By establishing clear boundaries, the overall well-being and behavior of the rabbits is improved. Colony housing is a suitable option not just for individual farmers, but also for larger operations such as rabbit barns or other related buildings. The design of the colony's dwelling is tailored to mimic a rabbit's natural habitat, allowing them to exhibit their natural behaviors and instincts. This type of housing system offers numerous benefits compared to traditional cages. It promotes socialization among rabbits, provides more living space for each individual rabbit, and encourages healthier levels of activity. Furthermore, this method also allows for easier monitoring and management of the rabbits by farm workers.

Colony housing in rabbit farming prioritizes the well-being of animals, offering advantages over traditional cages. It is increasingly popular among farmers aiming to create an optimal environment for their rabbits to thrive.

Fig. 9: Colony housing system

Rabbit Tractor Cages

Rabbit tractor cages are an essential tool for any rabbit farmer, providing a safe and secure environment for our furry friends while also allowing them to graze on fresh grass. These specially designed cages are built with the safety and well-being of rabbits in mind. The first important feature of a rabbit tractor cage is its strong frame. Constructed out of wire-covered steel or wood, these frames not only provide stability but also serve as a barrier to prevent rabbits from digging out or other predatory animals from entering. This ensures that our rabbits remain protected at all times. It's also crucial for the top of the cage to have a roof. This provides shelter for the rabbits during harsh weather conditions such as rain or excessive sunlight. The roof can be made from durable materials like metal or thick plastic to withstand any weather elements. To further enhance the weatherproofing of the cage, some farmers choose to add sheeting on the side panels. This helps seal off one corner and keeps rabbits dry and comfortable inside their tractor even during heavy rains. When it comes to size, we recommend using a 2 m by 1 m wide and 0.4 m high tractor that can hold up to 5 rabbits comfortably. It's important to adhere to the recommended space requirement of 0.4 square meters per rabbit to ensure they have enough room for movement and grazing. One unique feature of these rabbit tractor cages is their mobility. By relocating the cage every day, we allow our rabbits access to fresh grass while also keeping their living space clean and hygienic. This rotational grazing method not only benefits the health of our rabbits but also helps maintain healthy pastures. Rabbit tractor cages offer an excellent solution for housing rabbits in a safe, secure, and humane environment while providing them with daily access to fresh food sources. As conscientious farmers, it is our responsibility to ensure the well-being and care of our beloved bunnies in every possible way.

Fig. 10: Rabbit tractor cages

Nest Boxes

Nest boxes play a crucial role in the successful breeding and rearing of rabbits. Therefore, it is important to carefully consider their placement and design. On the 28th day following breeding, it is recommended to place the nest box inside the hutch. The ideal location for the box is in a hutch corner where a doe has been observed scratching or pretending to dig. This mimics her natural behavior and increases the likelihood of her accepting and using the nest box. However, it is important not to set up the nest box too early, as this may lead to unwanted behaviors such as using it as a toilet or consuming all of the nesting material inside. To prevent this, placing a small block or wedge under the front of the nest box can create an angle that encourages kindling at the back, while also protecting the kits from harm. The purpose of a nest box is to provide a safe and warm environment for both mother and offspring. It serves as a secluded space for kindling and protects against external elements that could harm or disturb the newborns. This concept stems from wild rabbit habits, where they create burrows within trees or ground for warmth and protection. While there isn't one perfect type of nest box that suits every situation, it is important that all boxes offer privacy for kindling as well as enough space for young rabbits to stay together without crowding. Careful consideration should be given to size and design when selecting a nest box, keeping in mind its intended use for nurturing healthy rabbit litters. Proper placement and design of nest boxes are essential factors in promoting successful breeding outcomes for rabbits. By considering their natural instincts and basic needs, we can provide them with an optimal environment for reproduction and growth.

1. Standard Nest Box

The Standard Nest Box is highly recommended for housing does during pregnancy and kindling. This type of nest box offers mobility, making it easier for the doe to choose a suitable location for her young. Typically, the doe will indicate her preferred nesting spot by scratching and collecting grasses or newspaper to create a comfortable nest. While there are various options for nesting materials, we strongly suggest using shredded newspaper as it is free from mites and other insects that can cause skin mange and ear canker. This will ensure the health and well-being of both the mother and her litter.

To allow ample time for the doe to prepare for kindling, we recommend placing the nest box in a designated area 25 days after breeding or one week before she is due to give birth. Additionally, it is important to slant the front of the nest box using a 2-inch block of wood to encourage the doe to kindle her litter at the back of the box. By providing your doe with a Standard Nest Box equipped with appropriate nesting material and proper placement, you are creating an ideal environment for successful kindling. We highly recommend this method as it has proven to be effective in promoting healthy offspring and increased litter survival rates.

To ensure the safety and well-being of the litter, it is important to take certain measures while constructing a wooden nest box. This includes cutting holes into the floor to keep the area dry and clean for the doe and her kits. However, it is crucial that these holes are not too large in diameter - preferably no bigger than a thumbtack's head - to prevent any harm to the kits. A small-gauge wire mesh can also be used on the floor of the box to allow waste to pass through easily. It is also important to consider heat loss if wind can enter from below, so proper insulation should be considered when building the hutch. Once the doe and her brood have finished using the nest box, it is essential to thoroughly clean, wash, and sanitize it for future use. These precautions will help ensure a safe and healthy environment for both mother and offspring.

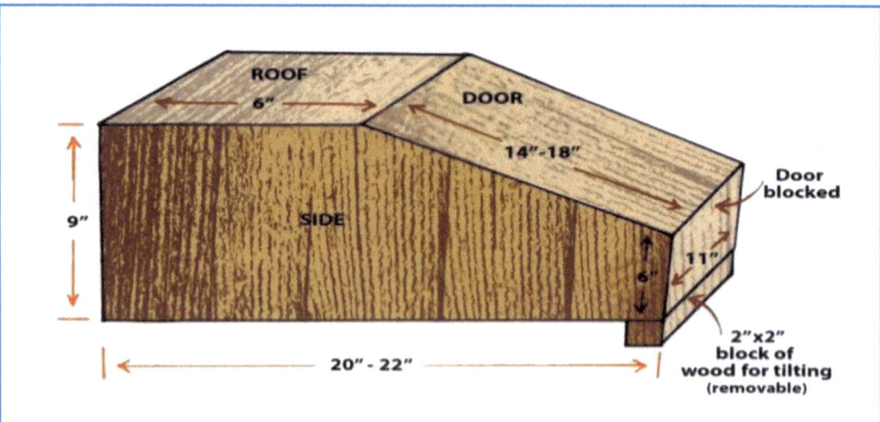

Fig. 11: Standard Nest Box

2. Counterset Nest Box

The counterset nest box is a commonly used nesting area for rabbits in the United States. It is built beneath the hutch floor, providing a natural burrowing space for the rabbits. These nest boxes are typically located at the front of the cage and have drawers that can be accessed from outside the hutch. One benefit of counterset nest boxes is that they create a more natural environment for rabbits, allowing them to exhibit their instinctual behavior of burrowing and seeking shelter underground. This setup also facilitates easier access for young rabbits, especially if they are separated from the nest at a young age. In traditional nest boxes, there is a risk of young rabbits jumping out and not being able to climb back in, which can lead to hunger or separation from their litter. This risk is higher in split litters, where different groups of young are kept in separate compartments. Therefore, counterset nest boxes can help prevent such issues and ensure proper care for all young rabbits.

Mother does typically care for their young in the nest or on the hutch floor during nighttime or in the early morning and evening hours, especially if they have been separated from their litter. The accessibility and convenience of counterset nest boxes make it easier for these maternal duties to be carried out effectively. Mother will not return the young to the nest or nurse both groups. The interior drawer of the counterset nest box can be easily removed for cleaning and disinfecting, simplifying maintenance. These drawers are also interchangeable among hutches. Once the younger offspring no longer need it, the inner drawer can be left out to create extra space in the hutch.

5

Feeding and Nutrition Requirements

In the rabbit production industry, providing a healthy diet and adequate amount of food every day is crucial for success. Rabbits can only survive and thrive on a diet of forage, so it is important to carefully consider their nutritional needs when planning their feeding regimen.

Some rabbit producers choose to provide basic grains in addition to forages in order to increase the potential productivity of their herd. These grains can be sourced from one's own farm or purchased commercially as concentrated feeds. The type of feed given to rabbits will greatly impact the management of the herd, such as how frequently they need to be rebred.

When it comes to raising rabbits, feed costs are one of the largest expenses. It is therefore essential to carefully consider the dietary needs of each individual herd in order to minimize costs and maximize productivity. This may involve choosing commercially prepared mixes or pellets, or creating a customized feed blend specifically tailored for your rabbits' needs.

By providing appropriate feeds and ensuring that rabbits are receiving optimal nutrition, producers can set themselves up for success in the rabbit production industry. Proper feeding practices not only contribute to the health and well-being of the animals but also play a significant role in overall herd management and profitability.

Feeding of Rabbits

Feeding is a crucial aspect of raising rabbits and requires careful consideration depending on the specific class of rabbits. Four groups have been identified based on their dietary needs: growing and fattening, resting does and bucks, pregnant does, and nursing does with litters. Each group requires a different amount and combination of feed to ensure optimal health and growth. It is essential to carefully monitor and adjust the ration supplied to each class to meet their specific needs. This ensures that the rabbits receive adequate nutrition to sustain their physiological demands without over or underfeeding them, leading to potential health issues. A well-balanced diet is key in supporting the growth, reproduction, and overall well-being of rabbits, making feeding an essential aspect for successful rabbit farming.

Feeding of rabbits is a crucial aspect of their overall health and well-being, with its "should bes" being predicated on the goal to have them perform at their best. These "should bes" include daily feed amounts and precise nutrient levels such as crude protein and Total Digestable Nutrients (TDN). Meeting these requirements is necessary to ensure the optimal growth, reproduction, and overall performance of rabbits. A commercial rabbit farmer would aim for a minimum of eight bunnies per litter, an average daily weight gain of 32 grams for growing and fattening rabbits, and a low rate of reproductive failure. These factors are indicative of ideal conditions that a skilled rabbit farmer would strive for in their operation.

Table 1: Rabbit Nutritional Requirements (based on air dry weight of ration)

Rabbit Class	**Body Weight (kg)**	**Total Feed Animal/day (gm)**	**Ton (%)**	**Crude Protein (%)**
Normal Growth Fattening Does/Bucks	1.8	114	65	-
	2.3	136	-	-
	2.7	155	65	16
	3.2	173	-	-
Maintenance Does or Bucks Resting	2.3	91	-	-
	4.5	150	55	12
	6.7	205	-	-
Pregnant Does	2.3	114	-	-
	4.5	186	58	15
	6.7	255	-	-
Lactating Does and litter of seven	2.3	-	-	-
	4.5	-	70	17
	6.7	-	-	-

Note: Approx daily gain is 32 grams (based on USNRC tables, 1966)

The following rules apply to a commercial-sized rabbitry that houses fifty or more animals:

1. Feed the rabbits up to 85 percent grass each day, or as much new grass and/or legumes as they desire.
2. Provide a commercial feed with the following values of crude protein, at least 16 percent:
   - 120 grammes per day for resting bucks and does
   - 240 grammes per day for pregnant women
   - Using litters works; 480 grammes per day

- Growing/fattening rabbits (when they are 3–4 months old or weaned for slaughter): 960 grammes daily

3. Always have fresh water on hand.
4. Trace mineralized salt (0.5 percent) should be included in rabbit diets.

Rabbits are commonly used as breeding stock for both their meat and fur. As such, it is crucial to ensure that the young produced are healthy and active. However, an imbalanced diet can lead to reduced reproductive performance and slower growth of the offspring. Concentrates, which are often used as a primary food source, have been identified as a major cause of this issue. It is important to note that for animals intended for breeding and meat production, using an inadequate or unbalanced diet cannot be justified due to the high costs associated with potential reproductive failures or smaller litters. It is therefore essential to carefully consider and monitor the diets of breeding rabbits in order to maximize their reproductive success and maintain a profitable output.

Table 2: Nutrient Requirements of Rabbits

Stages	Total (and Digestible) Protein %	Fat %	Fiber %	Digestible Carbohydrate % (NFE[a])	TDN %
Maintenance	12 (9)	1.5–2	14–20	40–45	50–60
Gestation	15 (11)	2–3	14–16	45–50	55–65
Lactation (with 7–8 litters)	17 (13)	2.5–3.5	12–14	45–50	65–75
Growth and finishing	16 (12)	2–4	14–16	45–50	60–70

[a] NFE = Nitrogen-free extract; TDN = Total Digestible Nutrient

Water Requirements

Water requirements for rabbits have been studied extensively, revealing that these animals have a higher daily water intake than dogs or cats of the same size. On average, a rabbit requires approximately 120 mL of water per kilogram of body weight each day. This finding is supported by the physiology and metabolic scaling of the rabbit's GI tract.

In cases where a rabbit is admitted to the hospital and is dehydrated, it is recommended to provide double the maintenance fluid dosage for at least 24 hours. This translates to 240 mL per kilogram of body weight per day, or 10 mL per kilogram per hour.

It is important to note that anorexic rabbits may also be dehydrated and will require additional fluids. Further research has shown that rabbits tend to consume more water from an open bowl compared to a sipper bottle. Therefore, it is advisable to offer water in an open container if possible.

Proper hydration is essential for maintaining the health and well-being of pet rabbits. It is crucial for owners and caregivers to ensure that their rabbits have access to clean and fresh water at all times. In cases where dehydration may be a concern, consulting with a veterinarian can help determine the appropriate amount of fluids needed for your specific rabbit's needs.

Carbohydrates for Rabbits

The role of carbohydrates in the diet of rabbits is a complex topic that requires careful consideration. While it may be tempting to simplify this aspect by stating that "carbs are bad for rabbits," the reality is not as straightforward. The impact of high-starch diets on adult rabbits is still a subject of debate, and their connection to dysbiosis predisposition remains uncertain.

Recent research has shed light on how adult rabbits metabolize starch differently from juveniles. This means that blanket statements about the harmful effects of starch on all rabbits may not always be accurate. However, it is essential to note that in young rabbits, certain types of polysaccharides (such as gluco-oligosaccharides) can lead to diarrhea, while others (such as galacto-oligosaccharides and fructo-oligosaccharides) do not have the same effect.

Fructo-oligosaccharides (found in fruits and vegetables) have been particularly beneficial for rabbit health. Studies have shown that diets supplemented with fructo-oligosaccharides can reduce morbidity in rabbits exposed to pathogenic bacteria such as Escherichia coli. As a result, many rabbit feeding products now include fructo-oligosaccharides as a critical ingredient.

It's also worth noting that growing rabbits can tolerate up to 15% molasses in their diet without negative effects. This ingredient provides not only carbohydrates but also essential nutrients like calcium, iron, and magnesium.

In conclusion, while there are valid concerns about the potential risks of high-carbohydrate diets for rabbits, the truth is more nuanced than simply labeling carbs as "bad." By carefully considering the different types of carbohydrates and their effects on different age groups and overall health, we can provide our furry friends with a well-rounded and nutritious diet.

Recommended Fiber Levels for Rabbits

Rabbits are herbivorous animals that require a high-fiber diet to maintain their overall health and well-being. The recommended fiber levels for rabbits have been carefully studied and established to ensure optimal nutrition and prevent potential health issues.

The role of digestible fiber in a rabbit's diet cannot be overstated. It serves multiple important functions, such as promoting dental health by providing the necessary chewing action, stimulating gut motility, increasing hunger and intake of cecotrophs (also known as "night feces" which help with digestion), and reducing behavioral issues like fur chewing.

For pet rabbits, it is generally recommended to provide a diet that includes up to 20% crude fiber and 12.5% indigestible fiber. However, it is essential to note that crude fiber measurements primarily capture the lignin and cellulose components of the food, which are not indicative of fermentable or easily digestible fibers.

Adequate levels of digestible fiber can be obtained from high-quality hay, leafy greens, and fresh vegetables. It is also crucial to gradually introduce new foods into a rabbit's diet to avoid digestive upset.

Maintaining the recommended levels of both crude and indigestible fibers is vital in preventing gastrointestinal issues such as diarrhea or constipation in rabbits. Providing a balanced diet with sufficient amounts of digestible fiber can contribute significantly to your rabbit's overall well-being. Always consult with a veterinarian for specific dietary recommendations for your pet based on its individual needs.

Concentrates

Concentrates, such as mash and pellet feeds, are essential components in the nutrition of rabbits. In order to find local sources for these concentrates, it is recommended to consult with a successful farmer or area extension agent. These individuals have experience and knowledge on what types of commercial feeds are available in the country.

There is a wide range of commercial mash and pellet feeds specifically designed for rabbits in the market. However, some farmers have also found success using other types of animal feed that are readily available. These include pig starter pellets, grower pig mush, and even pellets meant for chickens, pigeons, or other animals.

When choosing a concentrate for your rabbits, it is important to look for products that contain at least 16% crude protein. This will ensure that your rabbits receive enough energy to support their growth and development. It is also recommended to follow the feeding amounts suggested in the prescribed feeding regimen provided by the manufacturer.

In many developed countries, alfalfa hay makes up approximately half of the leguminous roughage included in pelleted rabbit meals. However, different regions may have varying types of roughage used in commercial rabbit feeds.

It should be noted that not all pellets or mashes designed for other animals will be suitable for rabbits. Therefore, it is important to carefully read labels and choose ones specifically formulated for rabbits to ensure proper nutrition.

By consulting with experienced farmers or area extension agents and conducting thorough research on available products, you can find high-quality concentrates that will help support the health and well-being of your rabbits.

Fig. 1: Pellets Feeding of Rabbits

Roughage, such as grasses and legumes, does not hold the same level of importance for animals other than rabbits. This is due to the fact that commercial feeds tend to consist mainly of concentrates, such as grains or grain byproducts. As a result, it is possible to reduce the recommended daily amount of concentrates provided by the US National Resource Council (USNRC) if ample amounts of roughage are included in the diet. However, it should be noted that the nutritional value of these forages is not equivalent to that of dried alfalfa used in commercial pellets. This is primarily due to their low digestibility and high water content. Therefore, reducing the amount of concentrate feeds can only be done in moderation.

Fig. 2: Feeding of Roughage

Due to the natural behavior of rabbits, they tend to waste feed when it is provided in a mash form. This is because they cannot easily claw out the pellets and end up squandering more feed. To counteract this issue, pellets are preferable as they are easier for rabbits to consume without causing irritation to their nose or lungs. However, if mash must be given, it is best to only wet the feed enough to prevent crustiness and limit waste. It is important to monitor the amount of moistened feed given and not leave it in the enclosure for prolonged periods as this can cause the feed to spoil and become unappetizing for the rabbit. In lowland areas where fermentation is more likely, it is recommended to provide rabbits with an adequate amount of concentrates so that they can finish their food within ten to fifteen minutes. It is crucial to carefully consider the amount of wasted feed in a large herd as even one kilogram per day can result in significant financial losses over time. Therefore, implementing strategies such as using pellets instead of mashes and controlling moisture levels in feed can help minimize waste and maximize efficiency in feeding rabbits.

Home Mixed Feed

A well-balanced and nutritious diet is essential for the health and well-being of rabbits. While commercial rabbit feed can provide essential nutrients, a home mixed feed can be a cost-effective and convenient alternative. However, it is important to carefully select and mix ingredients to ensure that all dietary requirements are met.

Firstly, it is important to note that rabbits can eat a variety of foods, as long as they are not greasy, sour, or spoiled. This includes items such as vegetable trimmings, fruit peelings, bread crusts, sun-dried leftover rice, and leftover milk. These can all be incorporated into the rabbit's diet in moderation.

The foundation of a rabbit's diet should consist of high-quality forage. This can include grasses and legumes such as alfalfa hay, timothy hay, clover hay, and fresh vegetables such as broccoli and dark leafy greens. Ideally, 85% of a rabbit's diet should come from forage.

In addition to forage, rabbits also require some commercial feed or homemade mixes to meet their nutritional needs. If these options are not available or preferred, green feeding consisting of 50% legumes and 50% other grasses can provide a minimal but adequate diet for dry bucks and does.

When creating a home mixed feed for rabbits, it is important to consider their specific nutritional needs at different stages of life. For example, pregnant or lactating does may require additional protein in their diet while growing rabbits may need more calcium.

It is recommended to consult with a veterinarian or experienced rabbit breeder when creating a home mixed feed to ensure that all necessary nutrients are included in appropriate proportions.

Numerous types of forages are suitable for feeding rabbits, as the table below illustrates.

Table 3: Feeds for Feeding Rabbits

Legumes	Grasses
Celery	Johnson grass
Bananas	Brome grass
Bean leaves	Pangola grass
Kudzu	Coastal Bermuda
Beet, sugar	Dallis grass
Cabbage	Tall fescue
Carrot roots	Bahia grass
Cassava, tubers	Kentucky bluegrass
Lettuce	Napier grass
Apples	Orchard grass
Comfrey	Bermuda grass
Sweet potatoes	Ryegrass
Kale	Sudan grass
Beet, mangel	Guinea grass
Cauliflower leaves	
Pea vines	
Potatoes	
Raisins	
Rape (colpa, colsa, colerape, tori, and chou oleifere)	
Rutabagas	
Dandelions	
Turnips	

Forages should be used in accordance with a few fundamental rules, just like other feed sources:

Don't Feed:

- Before feeding, don't pile up greens because they will get hot and ferment, which can lead to intestinal issues.

- The leaves or stems of the kamkong plant (which typically contain parasites prevalent in swampy environments).
- Forages gathered from areas where cats, dogs, and other animals frequently urinate since these areas may harbour tapeworm or coccidiosis.
- Sprayed forages or those recently exposed to pesticides

DO Feed:

- **Salt:** One teaspoon of salt should be added to the grain ration once a week or as needed. Salt can also be added to the feed at a rate of 0.5 percent or placed in the cage in a block or small container.
- **Water:** Rabbits must always have access to clean, fresh water. This is crucial, particularly in tropical regions where a doe and her litter might drink up to two litres of water every day.

  Finally, with regard to feeds, if you want to alter the rabbit's diet, go gradually. If animals used to just concentrate diet are given enormous quantities of lush greens, it could cause serious harm or even death. Digestive issues can arise from even little abrupt changes in a grain diet, which could be rather dangerous. To prevent any abrupt changes, the ideal approach is mixed feeding, which involves giving both commercial food and greens.

Coprophagy

Coprophagy is the act of consuming one's own excrement, a behavior commonly observed in rabbits. This behavior may raise concerns among new rabbit owners, especially if their hutch system has a wire-bottom which could potentially prevent it due to droppings falling through. However, these owners need not fret as rabbits are known to eat their cecotropes (soft fecal material) as they are expelled from the anus. This behavior typically occurs in the early morning when rabbits are less likely to be noticed.

Interestingly, this practice is often referred to as "pseudo-rumination" due to its similarity to the cud-chewing of ruminant animals such as cows and sheep. It is also worth noting that many rabbit breeders may not be aware of this natural behavior and may mistakenly view it as a sign of nutritional deficiency or inadequate feeding.

In reality, coprophagy is a normal and beneficial practice for rabbits. By consuming their cecotropes, rabbits can enhance the nutritional value of their diet. Therefore, rabbit owners should not discourage this behavior but rather understand and accept it as a natural part of their pet's digestive process..

Lactation

The last week of pregnancy is a time of fast mammary gland development. While it is possible for milk to be generated prior to kindling and even for high-producing does to leak milk from their glands, the actual letdown and production of milk often occur after kindling and are triggered by hormonal and neurological impulses from nursing. By the third week, milk production often reaches its peak and then starts to progressively decrease. The number of suckling young, the nutrition, and the amount of time the young are left with the doe all affect how long the breastfeeding lasts.

After the sixth or seventh week, milk production usually becomes insignificant; however, in well-fed, prolific does that have a litter size of eight or nine, milk production can continue for up to eight weeks. After the young are weaned from the doe at 8 weeks of age, milk is seen in their stomachs and continues to be secreted from the glands for a few days. Milk production is influenced by a number of variables, including food, genetic makeup, breed, and strain. According to a number of research on rabbit milk production, milk supply can reach 35 grammes per kilogramme of live weight at the peak of lactation. Based on this, the daily milk production of a 4-kilogram doe would be about 140 grammes (5 ounces). Despite what is commonly believed, the doe does not nurse her young for the whole twenty-four hours. Particularly for the very young in the nest box, nursing is typically done at night or in the early morning. It can just include a quick feeding that lasts a few minutes.

The young will attempt to nurse multiple times during the day once they have left the nest box and are eating solid food. But typically, the doe would ignore them and limit nursing to the evenings. Some does will, most rabbit breeders have seen, let the young nurse during the day. Animal behaviour scholars explain the doe's nursing practices by pointing out that rabbits in the wild are heavily preyed upon and unable to protect their offspring. The doe will therefore benefit from avoiding the young as much as possible. A doe's usual nursing schedule is determined by her regular feeding programme. Consequently, a doe should be able to consistently nurse her young between 6 a.m. and 7 p.m. if she is fed both concentrate and roughage at 5:30 a.m. and again at 5:30 p.m.

6

Management of the Herd

The management of a rabbit herd is a crucial aspect that requires careful planning and organization. As part of a larger scheme involving five medium-sized rabbit farms, the production and marketing of live rabbits and rabbit meat were streamlined into a single business. Although this may differ from the usual backyard production system, it serves as an ideal example of how one can modify their approach to suit specific needs. The success of this scheme lies in its detailed design, which ensures efficient management and optimal output for the herd. This includes proper breeding practices, regular health checks, adequate nutrition, and timely market analysis. With careful attention to these critical components, the rabbit herd can thrive in a profitable and sustainable manner.

Materials Needed

To enhance the organization and efficiency of your rabbitry operation, consider the following essential materials-

1. A large, promotional calendar from producers of feeds, seeds, and other agricultural inputs. These can often be obtained for free at agricultural sales outposts.
2. Colored dry-erase markers and a dry-erase board. The use of a white board is recommended as it allows for quick entry of new information and removal of old information. It also utilizes basic color coding for ease of use. This method is more efficient than using traditional pen and paper.
3. A notebook or binder to keep a permanent written record. This will enable you to track production over time and continuously improve your breeding stock. While the daily management system can function with just a white board, having a written record will provide valuable data for future reference.

These materials should be easily accessible in the rabbitry to ensure that tasks and activities are clearly visible and organized.

Guidelines

The whiteboard and calendar should be prominently displayed together at the farm, as both are essential tools for managing farm operations. To ensure the system's effectiveness, it is important to follow a set of core breeding guidelines. These guidelines can be taught, memorized, or provided in written form attached to the whiteboard and calendar. The following are the breeding guidelines that should be followed:

- Day 1: If the doe is receptive (swollen, darker vulva), bring her to the buck's cage.
- Day 14: Palpate her to check if mating was successful. If not, reevaluate receptivity the next day.
- Day 28: If pregnancy is confirmed after 14 days, set up a nest box in the hutch.
- Day 31: Three days after setting up the nest box, prepare for kindling.
- Day 33-60: Monitor for responsiveness for rebreeding, which can occur anywhere from a few days after kindling to two months, depending on nutrition quality.

Weaning Guidelines

The weaning guidelines for rabbits are a set of recommended steps to follow in order to ensure the health and well-being of both the mother doe and her kits. These guidelines have been carefully crafted by experts in rabbit care and should be strictly adhered to for optimal results.

Day 14: After 14 days from birth, it is important to remove the nest box from the doe's enclosure. This is because the kits should now be able to regulate their own body temperature and do not require the added warmth provided by the nest box. Removing the nest box will also encourage the kits to start exploring their surroundings and become more independent.

Day 28: Once the kits reach 28 days old, it is time to wean them from their mother's milk. This means separating them from their mother doe and introducing them to solid food. Weaning at this age is crucial as it allows for proper development of digestive and immune systems in the kits. It also gives the mother doe a chance to recover her strength before potentially breeding again.

During this process, it is important to monitor both the mother doe and her kits closely for any signs of distress or illness. It is also recommended to gradually introduce new food items into their diet over a period of several days, starting with small amounts before gradually increasing.

Use a grid of the farm (a picture of the arrangement or direction of the hutches) made on a white board based on the previously described "rules." Make each square in the grid represent a hutch. There must always be a minimum of one "task" every square containing a doe, denoted by a date written in the appropriate colour for that particular kind of task. The assignments match the above-mentioned guidelines, as does the quantity of coloured markers required. The fundamental idea is that feeding and watering the cattle involves the first trip through the farm as part of the daily tasks. A simple inspection of the hutches during this process can identify any unusual problems that require attention. The farmer looks at the white work board after watering and feeding. He or she looks for the date for that day, as well as any boxes containing past-due dates. The farmer may determine which hutch has a work to be accomplished (based on the grid that depicts the farm's layout) and what the task is (based on the colour of marker used) by looking for the date that day (or a circled date that has past).

Fig. 1: A visual depiction of weaning and breeding is created by a rabbit breeder

The foundation of the system is the idea that there is always a new work to perform when a previous one is finished. Therefore, the farmer should note the following day for which a job is needed if a doe is successfully palpated (on Day 14). After 14 days (Day 28) in the previous example, the next step is to install a nest box inside the hutch. The calendar with the white board shown is necessary for efficient usage of the system since it is necessary to write in future chores. In a given grid square, there is often only one duty assigned, though occasionally there are several. For instance, the doe will be given a date after kindling on which she is to be watched for receptivity and on which day the kits are to be weaned. If additional cages are kept for young fryer (non-producing) rabbits, such cages can also be mapped onto the grid; however, as they are not employed for "management tasks," they can either be blacked out or have a "X" placed on the corresponding grid square. The farmer circles the date in the same colour as the task if a doe is not found to be responsive on the day it becomes necessary. Tasks that are past due are indicated on the board in this manner so they can be reviewed the next day. An independent, yet connected, management tool is a written record kept in a file or folder. This offers a permanent record of the herd, unlike the white board. Farmers can track the productivity of the breeding females in the herd and start selection based on the productivity of their breeding stock by keeping production sheets on does. It is more accurate to evaluate the number of successfully weaned kits than the number of kits born for assessing productivity.

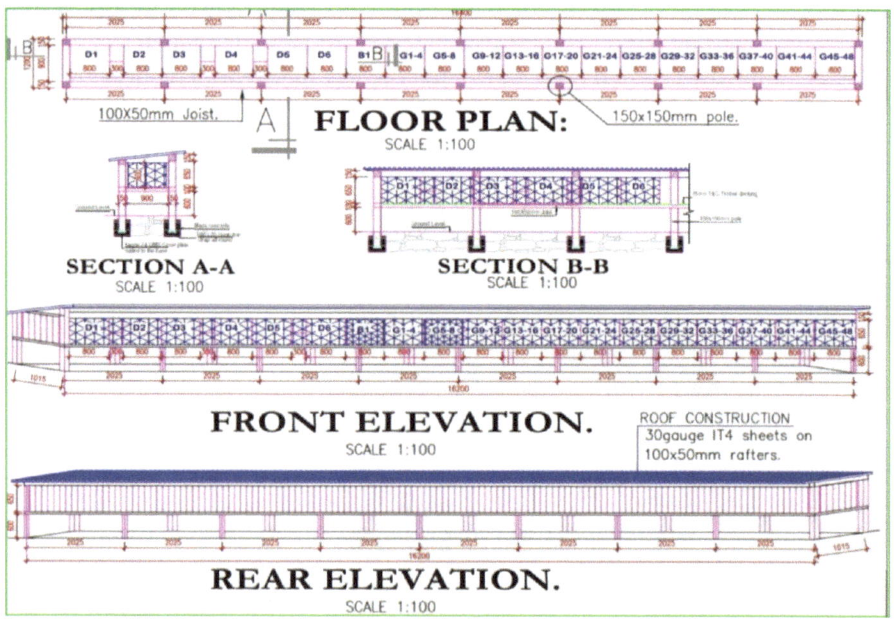

Fig. 2: A proposal for an outdoor hutch that will house six does and one buck in an economy unit

Ways to Care for Rabbits

Never pick up a rabbit by its legs or ears. This kind of handling could harm them and perhaps result in drooping ears. Can lift a rabbit without risk if use the scruff on the back of its neck. Larger animals should be supported with a second hand underneath the bottom. A rabbit can be turned to appear to be lying with its back on the holder's forearm while it is grasping onto the scruff. This will often subdue a large number of rabbits and is helpful for assessing overall health, including the state of paws or mammary glands, or to identify the sex of the animal or determine whether a female is receptive to sexual stimuli. Once a rabbit has been picked up by the scruff, the best way to carry it is to nestle its head into the crook of the opposing elbow and use the palm of that hand to support its rear legs and bottom. It's similar to the method used to teach football carrying to American football players.

Small rabbits can be lifted and carried easily by softly and firmly grabbing the loin area. Point the hand's heel in the direction of the animal's tail. By using this technique, the pelt won't be harmed or the carcass bruised. A medium-weight rabbit can be lifted and carried by letting the right hand stroke down on the ears and grabbing the skin fold over the shoulder. This gives the head section more control. Put the left hand under the rabbit's rump to support it.

Fig. 3: Little rabbits can be lifted by gently grabbing the loin area, as demonstrated here

Use a similar technique to lift and carry heavier bunnies. If the rabbit tries to scratch and struggles, release it on the ground and repeat the above instructions, or tuck its hind legs under the right arm and hold it tightly.

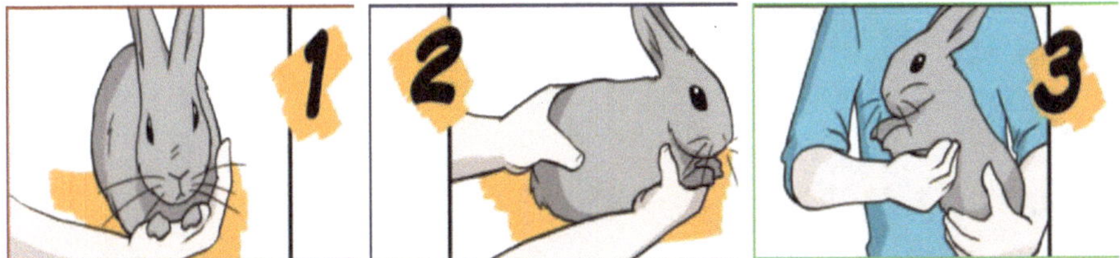

Fig. 4: Tuck the rear legs under the right arm and hold tightly to carry heavier rabbits, as demonstrated above

Using Palpation to Assess Pregnancy

Palpation is arguably the most important and useful "hands on" skill that a beginner rabbit producer needs. The main idea is that you can pretty much tell if a rabbit is going to kindle or not right before the halfway point of a potential pregnancy. This is crucial in a productive system because it prevents time wastage, and time is money.

A doe can be palpated 14 days after the buck services her, and neither she nor the foetuses are in any real danger. The foetuses are sufficiently submerged in a protective fluid at this point in development, thus palpating them at a modest pressure is safe. Palpating presents a higher risk of injury later in development and before to kindling. When the doe reaches 14 days, a knowledgeable farmer can decide that she is not pregnant and start looking for receptivity indications again, such as an enlarged, reddish-purple vulva, before putting her back in the buck's cage for service, at which point the cycle repeats itself. The method is difficult at first, but it becomes a readily used approach very fast. The doe can be restrained by using one hand to grasp the skin behind its ears and neck and the other to reach below its body, as demonstrated in the photo below. This second hand may move smoothly down the abdomen towards the lower area, almost in between the hind legs, with an open palm.

A nervous doe is stressful for the animal and difficult to palpate adequately, so wait for the doe to calm down. The thumb and first two fingers should start searching the belly with the palm facing up and the hand in contact with the abdomen, looking for round or rectangular bulbs about the size of a large grape (smaller is palpating early than 14 days). Palpating involves applying slightly more pressure while bringing the fingers and thumb together and relaxing in order to feel for contents inside the actual abdominal cavity. It is not as simple as lightly massaging the skin. The exercise should end and the doe put back in her hutch as soon as the palpation reveals the existence of foetuses.

Fig. 5: The right approach to palpate the doe is demonstrated

A skilled palpator can impart the skill to others through practical demonstrations. The Peace Corps Volunteer is passing along the knowledge to a counterpart, as seen in the picture above. The doe should be placed on top of the hutch, on a table, or on any other level surface by the teacher, who should then place the first hand softly but firmly on the doe's back of the neck to restrain her. The student of the technique should then insert his or her hand beneath the instructor's hand exactly as they are going to place the second hand beneath the tummy, palm up. The student's right hand must be positioned beneath the instructor's upward-facing palm if the instructor is palpating with their right hand. Subsequently, the pupil must slip his or her hand forward slightly so that the thumb and first finger are visible while the remainder of the hand is usually buried beneath the instructor's hand, prior to the instructor moving this second hand beneath the doe. The palpation can then be performed as previously stated by the instructor as they slowly slide the clasped hands beneath the doe. The distinction is that the instructor may find the foetuses and transfer them from between their own thumb and forefinger to the student's thumb and forefinger by placing the second hand beneath and slightly ahead. For this to work, the student's thumb and forefinger must be below the instructor's and closer to the doe's tail, and they must also "track" the instructor's finger movements.

A beginner farmer of rabbits should presume that a doe is pregnant and handle her appropriately if they are unable to palpate the animal correctly. While there are other possible indicators of pregnancy, none are quite as reliable as pregnant palpation. Returning the doe for service after mistakenly believing that there isn't a pregnancy can lead to issues because it's conceivable for her to conceive a second litter concurrently with the first. A doe's reproductive system consists of two "horns," each of which is capable of supporting a litter by itself. Even while double litters are uncommon, they are not as uncommon as two successfully weaned litters that were born within days or weeks apart.

Kindling

The actions listed below will guarantee successful kindling:

1. About 28 days after the doe mates, place a nest box in her favourite part of the hutch. This guarantees a suitable location for the young to be born and enables the doe to build a nest ahead of time.
2. They occasionally forget to cover their litter with fur, or they ignite the litter on the hutch floor and allow it to get cold. Even if the young seem dead, you might be able to save them by warming them if you find them in time. Heat up a cup of lukewarm water for the babies. Take the infant by the head and dip it into the cup several times. Then use a cloth to dry. After the young have warmed up, arrange the bedding to form a cosy nest. From then, the doe usually takes over.
3. It is simple to remove the doe's fur for kindling, and you can remove enough from her body to cover the nest's litter. Having extra fur on hand is a good idea in such situations. In order to maintain the nest clean, remove some hair from areas where does have pulled too much and

store it nearby in a different bag or box. While sterilising or deodorising the fur is not required, you should take extra precautions to keep the doe from smelling it. Hold the odd fur about her legs and run your hand down her back before putting the fur in the nest box. This helps her scent travel to the unfamiliar fur. If these precautions are not taken, there's a good chance the doe will eat the unfamiliar fur if she scents it.

4. The doe often takes less food as usual a day or two prior to kindling. Try to make her feel as comfortable as you can without disturbing her. At that point, you could entice her with modest amounts of commercial and green feed. Her digestive system is going to benefit from this. Give the doe an abundance of fresh green feed once she kindles.
5. The majority of litters are lighted at night. Upon igniting, the doe might become agitated. Wait till she has calmed down before disturbing her.

Young Litter Care

When they are 19 or 20 days old, young rabbits begin to emerge from the nest to eat food. They open their eyes around 10 days. The young may not be receiving enough milk, the nest may be overheated, or the door blocker at the entrance of the nest box may be positioned too low if they leave the nest box too soon. During the first two weeks of nursing, the doe typically feeds her young at night or in the early evening and morning. She will nurse them at her discretion after two weeks. The doe will nurse the young in the nest or those on the floor if the litter splits up. She will not take the young and put them back in the nest; she will not nurse both groups. This happens if the nest box is too big or isn't inclined upward or downward at the front.

After lighting the fire, take the nest box out of the cage and go over the litter, making sure that there are no dead, malformed, or undersized eggs. In most cases, the doe won't protest if you examine it quietly and carefully. She won't be in any risk of rejecting the young. To calm her down and divert her attention if she is agitated and anxious, put some enticing food in the hutch right before inspection. The size of litters varies. Typically, the more popular breeds have eight young on average. Many producers only produce eight kits every litter since does normally only have eight teats. The worry is that only eight kits can eat at a time, which means litters of nine or more are at a feeding disadvantage because the doe only offers the chance to feed once a day. Kits will have a better chance of surviving if they are fostered by a different mother who has a litter of about the same age. Recall that the quantity of puppies who are successfully weaned matters considerably more on a productive farm than the size of the litter at kindling.

A foster mother with a small litter of rabbits may receive some of the newborn bunnies from a large litter. At the time of weaning, more uniform growth and development are ensured when the quantity of young is adjusted to the doe's capabilities. As a result, if multiple mates do this, they will light up about simultaneously. The young that are transferred should be three or four days older than the foster mother's young in order to yield the optimum results. If the kittens are in good health and near in age to the adoptive litter (not already cold or otherwise obviously declining), fostering is often safe and successful.

Reasons for Losses in Infant Litters

1. Loss of milk

One of the most common reasons for losses in Raabit infant litters is the lack of milk production by the mother doe. This can happen due to a variety of reasons, such as inadequate nutrition or health issues. When the problem is not identified and addressed promptly, it can lead to severe consequences for the young rabbits. If they are placed with foster mothers, they may still not receive enough nourishment and eventually starve to death within a few days. It is essential for rabbit breeders to closely monitor newborn litters in the first few days after delivery to ensure that they are receiving proper care and attention from their mother. Any signs of inadequate milk production should be immediately addressed to prevent any potential losses in the litter.

2. Eating young

One of the reasons for losses in raabit infant litters is cannibalism by the mother doe. This could be due to various factors such as inadequate or poor quality ration, anxiety or lack of parental instincts in certain strains of rabbits. The cannibalism is usually limited to dead or injured young and not healthy ones. While it may be disheartening for producers, especially new ones, it is important to give the doe another chance as many eventually become excellent producers after their first litter. It is crucial for rabbit producers to closely monitor their doe's behavior and provide necessary support and resources to ensure a successful litter rearing process.

3. Disturbance

One of the main reasons for losses in raabit infant litters is due to disturbance. The doe, or mother rabbit, may become frightened and start to kindle, or give birth, on the hutch floor if she is disturbed. This can lead to exposure and result in the litter perishing. Predators such as cats, snakes, dogs, and ants may also cause disturbance for the doe even if they cannot enter the rabbitry itself. In addition, if the caretaker's presence is unfamiliar or sudden during the kindling process, this can also startle the doe and potentially cause her to harm or kill her litter by stomping with her back feet. To prevent this from happening, it is important to keep strangers away from the rabbitry during breeding season and only allow familiar caretakers in order to minimize any potential disturbances that could result in loss of a litter.

4. Weaning

Weaning is a crucial stage in rabbit production, involving transitioning young rabbits from mother's milk to solid food. The duration of weaning depends on the animals' nutritional intake. Ideally, baby bunnies can stay with their mother for 8-10 weeks in a system providing high-quality feed.

As the young rabbits consume more solid food, the doe naturally produces less milk, allowing her to rest and prepare for the next breeding season. By 14-16 weeks, fryer bunnies should reach a marketable size and weight.

During weaning, it's essential to gradually introduce solid food to prevent health issues and stunted growth. Proper nutrition is crucial for the rabbits to thrive and reach their full potential for future production.

Careful attention and monitoring are necessary during weaning to ensure the rabbits' health and success. Adjustments should be made based on individual needs. Successful weaning leads to healthy, marketable rabbits ready for consumption or breeding purposes.

Determining the Rabbits' Gender

Determining the gender of rabbits is an important step in managing a breeding program or selecting replacements for your herd. It is recommended to separate the sexes at weaning, which can be done accurately when the baby rabbits are less than a week old, but it becomes simpler after they have been weaned at eight weeks.

To determine the sex of a rabbit, it should be gently but firmly restrained to prevent it from moving around. One common method is to hold the rabbit with its head up and legs between your fingers, then use your left hand to secure its chest while using your right hand to depress the area in front of its sex organs. This will reveal the crimson mucous membrane. Another technique is to place your thumb under its right hind leg and use your index and forefinger to push down on the area.

If you are not confident in using these techniques, another way to determine gender is by observing their behavior while eating. If male rabbits are already mounting or riding each other at one and a half months old, then they can be considered potential breeding males for future use.

It is important to accurately determine the gender of rabbits for proper management of breeding programs and selecting replacements. By following these methods, you can ensure that you have correct information about the genders of your rabbits for successful breeding and production in the future.

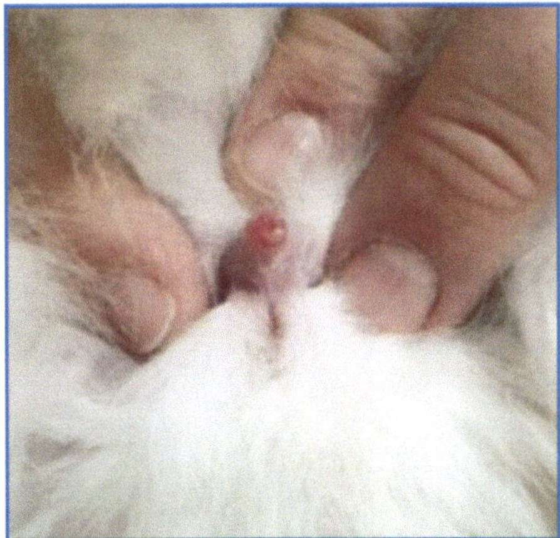

Fig. 6: Sex organ of a buck

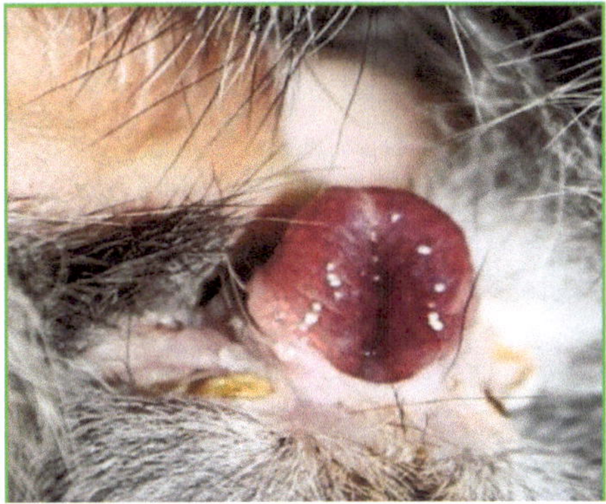

Fig. 7: Sex organ of a doe

Record Keeping

Maintaining records is crucial to effective management. The above cards highlight the key components of a basic record-keeping system. Make a note of every breeding rabbit for your database. The greatest marking technique is getting a tattoo because it is non-disfiguring and permanent. For tattooing, it's convenient to confine rabbits in an adjustable box. Due to their tendency to pull out and deform the ears, ear tags and clips are not a suitable marking option.

HUTCH RECORD CARD

Doe ________ Ear Tattoo No. ________ Breed ________ Normal Weight ________

No. Nipples ________ Born ________ Sire ________ Dam ________

Buck	Date of Service	Pregnancy Check	Date Kindled	Number of Young		At 8 weeks		Remarks
				Kindled	Left	No. Left	Weight	

Fig. 8: An Example of a Record-Keeping Hutch Card

Keeping track of breeding, kindling, and weaning procedures requires an easy-to-use record system. Desirable breeding stock can be chosen and ineffective animals can be culled using information from the records.

BUCK RECORD CARD

Name: ____________ Ear #: ____________ Breed: ____________

Sire: ____________ Dam: ____________ Birthdate: ____________ Weight: ____________

Doe Served	Service Date	Kindle Date	Number of Young				Average Wt. (3 weeks)	Average Wt. (8 weeks)	Notes
			Born	Alive	3 wks	8 wks			

Fig. 9: An Example of a Record-Keeping Buck Card

Unless the operator's schedule and personal preferences dictate otherwise, records do not need to be very extensive. Any documentation maintained must enable the operator to determine production expenses and assess the level of advancement over similar time periods.

The number of does that bred, the number of conceptions, the number of does kindling, the number of does raising a litter, the total number of young left with the doe, and the total number of young weaned or reared each breeding are among the basic details that are sought. These information will supply the ongoing manufacturing components that are required. A monthly summary form can be created by compiling data from the hutch record cards.

DOE - *Breeding History*

Doe: Name | Number | Date of Birth | Hutch Number | Breed

Parents: Sire: | Dam:

History

Mating Date	Buck	Nestbox Date	Kindling Date	Number of Babies on Day:					Weaning Date	Bucks	Does	Remarks
				1	2	5	7	14				

Fig. 10: An Example of a Record-Keeping Doe Breeding History

The monthly data can then be added up to a yearly summary form. By displaying the total investment, income, and expense numbers on a summary chart, one can determine the annual summary of the rabbitry. Keeping track of records is crucial for success in any kind of rabbitry, be it small-scale, backyard, or commercial. From historical records, highly productive does and bucks can be obtained for replacement stock and sale.

Rabbit Worksheet

Name____

Year________________Grade in school ____________________

My project began with the following animals:

Date	Description	Beginning Cost	Cost or Value
		Total Weight (Box 1)	Total Cost (Box 2)

During the year, the following animals died:

Date	Description	Estimated Value	Cause

Iowa 4-H Rabbit Worksheet Feb-18

Fig. 11: Example of a Rabbit Worksheet

Preventing Accidents

Preventing injuries is a part of sound management, which involves considering the wellbeing of the animals. Many injuries, like paralysed hindquarters in rabbits, are typically the consequence of mishandling or slipping in the hutch when exercising or trying to flee from predators. These slips typically happen at night, around the time of kindling. Damage from defective cages with protruding wire, nails, or incorrect wire size is another source of injury. Dislocated vertebrae, injury to nerve tissue, and strained muscles and/or tendons are common injuries.

The animal might heal in a few days if the damage is not too severe. Provide a healthy nutrition and comfort to the injured animal. To save needless misery, destroy it if it doesn't get better in a week. Therefore, it's critical to provide your bunnies a peaceful, pleasant environment as well as protection from predators and unneeded disruptions. Emphasize once more that noise should never be made in the rabbitry. Additionally, refrain from allowing guests to prod the bunnies; instead, treat them with courtesy.

Another precaution is to have your toenails trimmed. Rabbits kept in hutches develop abnormally long toenails. They might even grow to a length where they distort the foot. Additionally, the nails could snag on the wire mesh floor, inflicting pain and misery. Cut the nails on a regular basis with side-cutting pliers. Make a cut in the toenail below the point of the cone. If the foot is held up against the light, the cone can be seen. There won't be any bleeding or harm to the delicate area from this.

Hygiene and Disease Control

The simplest approach to guarantee hygienic conditions and prevent illness is to clean hutches, containers, and the surrounding area on a daily basis. You can significantly reduce the risk of contracting any disease in the rabbitry or from getting sick yourself by following strict sanitation measures like cleaning all cages and water containers on a daily basis and gathering roughage from uncontaminated areas (ideally, you will have allowed space for planting your own forages). Those interested in a medical approach to illness management are recommended to carefully explore smaller production systems or backyard rabbit husbandry. Even while some veterinary therapies can be very beneficial, determining the need for them also requires access, money, and expertise. To put it simply, a lot of producers—even those operating on a somewhat larger scale—manage their herds well and keep most diseases at bay by keeping their rabbitries tidy and hygienic. Hutches, nest boxes, and crocks or other containers holding food and water should all be cleaned out. Furthermore, this type of management includes making sure that there is adequate ventilation and periodically removing hair accumulations from hutches. The farmer raising rabbits can prevent and manage disease by using ventilation.

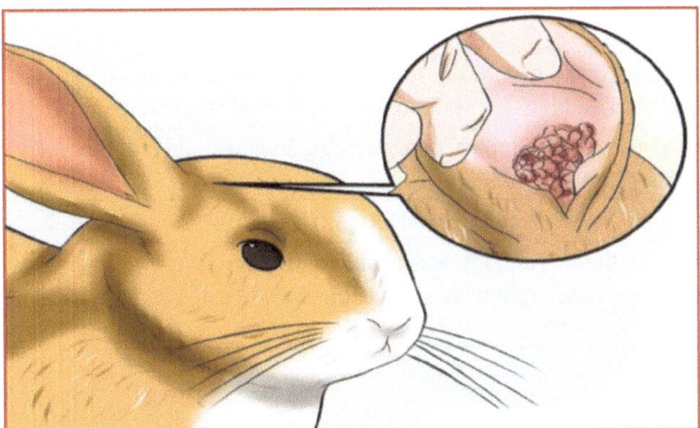

Fig. 12: An ear mite infestation causes a rabbit to maintain an asymmetrical head position

The rabbit is an extremely prolific animal. There's no doubt that the relative fragility of the animal (especially young animals) has contributed to this evolutionary feature. It is imperative to acknowledge the existence of sickness and mortality on the farm in order to effectively oversee the system. With this understanding, the most effective, affordable, and even ethical method of treating illness may involve a systematic strategy to culling sick animals. An animal that has been culled owing to disease should generally be destroyed and declared unfit for human consumption, unless it is known to be safe.

Table 1: Prevention of Diseases

Status	Prevention
Vent disease	Best to cull
Ear mange-Ear mites	Keep a small spray bottle filled with baby oil; spray into ears to treat or even prevent
Skin mange	Check with local extension agents for information
Sore hocks	Choose high-quality flooring materials. Steer clear of jagged or rough edges, fix cracks, and other flaws that could rip or abrade skin.
Warbles	Check with local extension agents for information
Mastitis	Best to cull
Urine-hutch burn	Keep the rabbitry clean by keeping litter or urine from collecting in the corners of the cages.
Ringworm	Best to cull
Pneumonia	Can be prevented or minimised with a hygienic environment, but can be isolated first, monitored, and culled if no improvement
Infected nose	Check with local extension agents for information
Caked breasts	Wean kits gradually; for example, if there are eight kittens in a litter, wean two of them on the first day, two on the third, two on the fifth, and the last two on the seventh.

Lympjadenitis	Best to cull
Snuffles	Can be prevented or minimised with a hygienic environment, but can be isolated first, monitored, and culled if no improvement
Weepy eye	Keep a strict eye on things, separate sick animals, and cull them if they don't go away.
Bloat	For details, speak with the local extension agents.
Heat prostration	Make sure there is enough water and ventilation. When stressed or left without water, rabbits become extremely hot. Particularly while they are transporting.
Fur block	Check with local extension agents for information
Buck teeth	Cull
Pinworms	For details, speak with the local extension agents.
Pseudotuberculosis	Best to cull
Metritis	Check with local extension agents for information
Coccidiosis	For details, speak with the local extension agents.
Tapeworm	Check with local extension agents for information
Wry neck	Best to cull—can prevent/limit by managing ear mites effectively
Papilloma	Check with local extension agents for information
Listeriousis	Best to cull
Milkweed poisoning	For details, speak with the local extension agents.
Hydrocephalus	Best to cull
Paralyzed hind quarters	Best to cull

Fig. 13: Manure gathers in the cage's corners every day. Skin mites grow and infest the rabbit if it is not cleansed on a regular basis

Other recommendations for illness prevention are as follows:

1. In a way, illness is a natural occurrence that can never totally be eradicated but can be significantly reduced with a strict daily hygiene regimen.
2. Sanitation procedures that are intentional and thoughtful can typically keep sickness at a minimum.
3. Keep animals apart from one another. The best course of action is prevention rather than therapy or even a potential cure; good hygiene habits are PREVENTION.
4. While not always in such an obvious pattern, characteristics like strong natural resistance, long life, and high output are undoubtedly inherited in the same way as other qualities like size, colour, ear length, etc. Consistently choosing breeding stock based on exceptional performance will be well worth the effort.
5. Adhere to a healthy diet in order to maximise the expression of your best inherited characteristics.
6. Offer lots of airflow that is free from drafts. When the edges of a self-cleaning floor are overly enclosed, updrafts occur, which are very unpleasant.
7. Allow the animals to receive enough of sunlight, even in the absence of intense heat. There must be shade as well.
8. Maintain all equipment in good repair to reduce the chance of accidents and keep it CLEAN and DRY.
9. Refrain from handling animals, feed, water and food containers, or any equipment they come into contact with needlessly. The caregiver's hands and clothes have the potential to transmit illness.
10. Separate animals that may be infected with infectious diseases, and tend to them after the healthy animals have received care.
11. Whether an animal is being added to your herd or one of your own that may have come into touch with other rabbits-either directly or via handlers and equipment-keep it apart for a period of one to two weeks.
12. Keep the animals safe from any unsettling influences, especially nocturnal scavengers. As long as they receive regular care, let the animals spend the entire day sleeping.
13. Keep marketable stock separated and confined outside the rabbitry or at the entry if rabbits are regularly sold to a dealer. The person who picks up the rabbits will be grateful for your assistance in reducing the likelihood that they may contribute to the spread of illness, as they visit numerous rabbitries in quick succession.

Habit of 'Fur' Eating

The act of rabbits consuming their own fur, as well as fur from bedding or other rabbits, is often a red flag for inadequate nutrition in their diet. Lack of sufficient bulk and fiber are common causes for this behavior. In some cases, a shortage of protein may also be the culprit. To address this issue, it is recommended to increase the intake of legumes such as soybean, sorghum, or peanut meal.

A responsible breeder closely monitors the overall health of each rabbit in their herd and makes necessary adjustments to their diet accordingly. This may involve customizing the amount of food given to meet each rabbit's specific nutritional needs. Additionally, providing high-quality grass, fresh leguminous feed, or root crops alongside home-mix or pelleted food can help curb abnormal eating habits.

Alternatively, another solution is to refrain from feeding the rabbits for one full day. This will allow them to reset their appetite and encourage them to consume a more balanced diet moving forward. Overall, maintaining a well-rounded and nutritious diet is crucial for promoting healthy behaviors in rabbits and preventing them from resorting to consuming their own fur.

Avoiding 'Fur' Block

Rabbits have a natural grooming behavior in which they lick their coats to clean themselves. However, this can lead to the ingestion of fur or wool that is not properly broken down, potentially resulting in a dangerous condition known as "fur block". This occurs when a large amount of fur accumulates in the rabbit's stomach and obstructs digestion, ultimately leading to starvation if left untreated.

To prevent this, regular shearing of Angora rabbits is highly recommended. By removing excess fur from the animal's coat, the risk of fur block can be significantly reduced. Additionally, providing rabbits with sufficient protein and roughage can also discourage them from consuming their own fur.

In order to reduce the likelihood of rabbits chewing on their own fur, it is beneficial to provide them with a block of wood that has been soaked in salt for three days. This will not only satisfy their natural instinct to gnaw but also deter them from consuming their own fur.

Overall, it is important for rabbit owners to be aware of and take precautions against fur block by regularly shearing Angoras and providing appropriate nutrition and environmental enrichment for their pets. By following these measures, we can help ensure the health and well-being of our beloved rabbits.

Gnawing the Hutch's Wooden Parts

Rabbits have a natural tendency to gnaw on wood, as it provides them with essential nutrients and helps keep their teeth trimmed. However, excessive gnawing can be a sign of inadequate salt intake. If your rabbit is excessively gnawing on the wooden parts of their hutch, it is important to address this behavior.

When constructing the hutch, it is recommended to cover any exposed wooden edges with wire mesh or use materials such as flattened cans, galvanized iron, or tin strips. It is crucial to ensure that these materials do not have sharp edges that could potentially harm your rabbit.

In addition to covering exposed wooden areas, providing a block of wood soaked in brine solution can also help deter rabbits from gnawing on their hutches. This will fulfill their need for salt and discourage them from damaging the hutch.

Furthermore, offering high-quality grass and incorporating legumes and root vegetables into your rabbit's diet can decrease their urge to nibble on wooden structures. Ensuring that rabbits have access to a well-balanced diet can greatly reduce their need to chew on inappropriate items.

In summary, while it is natural for rabbits to gnaw on wood, excessive gnawing may indicate a deficiency in nutrition or boredom. Implementing these preventative measures such as covering exposed wooden edges and providing appropriate alternatives can help alleviate this behaviour. As always, consult with a veterinarian if you have concerns about your rabbit's chewing habits.

Ideal Environment for Rabbit Rearing:

- **Lighting:** Lighting plays a vital role in creating an optimal environment for rabbit breeding. Both natural and artificial light are necessary for successful reproduction. Bucks need 8 to 12 hours of light daily for proper reproductive function, while does should have at least 6 hours of light exposure for optimal fertility.

  To provide suitable lighting, a combination of natural and artificial sources is recommended. A 40-watt fluorescent tube or a 100-watt light bulb placed two meters above the ground and spaced three meters apart can be used. These lights should be on for 16 hours from 6 a.m. to 8 p.m. to mimic natural daylight. Sudden changes in lighting should be avoided to prevent stress and injuries. Young rabbits require 1 to 2 hours of light exposure daily for growth and development, unlike adults who need longer periods of light.

- **Temperature:** Maintaining the ideal temperature is crucial for the health of rabbits. They are most comfortable between 10°C to 26°C, although they can tolerate temperatures from 5°C to 33°C. In India, winter temperatures are generally suitable for rabbit rearing, except in hilly areas. Rabbits handle cold better than heat. During summer, prevent heat stress by ensuring proper ventilation and cooling in their living space. Avoid drafts that can cause discomfort. Adult rabbits can regulate their body temperature by stretching out or curling up, but newborns may struggle with sudden temperature changes, leading to potential death if not monitored.

  It is vital to maintain a consistent temperature for rabbits to thrive and stay healthy. Keep a close eye on their environment to ensure their well-being.

- **Humidity:** The ideal environment for rabbit rearing requires careful attention to the humidity levels within their living space. It is vital to maintain a humidity level of no more than 50% to ensure the rabbits' well-being. During the rainy season, all necessary measures must be taken to reduce humidity, such as utilizing appropriate equipment. High humidity combined with high temperatures can be detrimental to rabbit health and should be avoided at all costs. Proper maintenance of irrigation systems is crucial in preventing water leaks, which can increase humidity levels and pose a risk to the rabbits' health. Additionally, storing water bottles outside can help prevent breakage and further decrease humidity levels within the rabbit house. By maintaining optimum humidity levels, we can provide an ideal environment for rabbits to thrive in our care.
- **Ventilation:** An ideal environment for rabbit rearing requires proper ventilation to ensure that the rabbits can breathe freely and live in a clean and healthy habitat. The area where the rabbits are housed should be free from smoke and dirt to promote good air quality. It is crucial to create an unrestricted airflow by strategically placing the ventilation outlets in the right locations within the home. During hot summer days, it is essential to provide access to warm, fresh air while avoiding strong drafts. Additionally, providing shade-giving trees around the rabbitry can help regulate the temperature and let cool air into their living space. By maintaining optimal ventilation conditions, we can provide a comfortable and conducive environment for our rabbits' well-being.
- **Noise:** An ideal environment for rearing rabbits should prioritize minimizing noise pollution. While there is limited research on the direct effects of sound on these animals, it is widely known that loud noises

can disrupt their innate behaviors and natural rhythms. This interference in turn can have negative impacts on important aspects such as breeding and maternal instincts, ultimately affecting the overall health and well-being of the rabbits. Therefore, in order to create an optimal environment for rabbit rearing, it is essential to minimize any excessive or disruptive noise levels. This includes selecting a suitable location away from noisy areas and limiting any loud activities or machinery near the rabbits' living space. By prioritizing a quiet environment, we can ensure that our rabbits are able to thrive and reach their full potential without unnecessary disturbances.

7

Management of Rabbit Farm Reproduction

There are three sizes of rabbit farms: small, medium, and giant. Initially, their owners typically had a small number of does and a stock of growing bunnies for their personal use. Rabbit farms began expanding and selling animals when butchers began to sell rabbit meat. Breeds, housing conditions, feeding practices, and management strategies were all developed in accordance with the increase in does. In Europe, backyard stock owners have a mediocre interest in expertise. The majority raise rabbits in the same manner as their parents or grandparents, adhering to family custom. Medium-sized farm owners are eager to learn more and make the most of their meagre financial resources by using this knowledge. Farmers grow their farms larger in order to reap greater benefits because profit per animal or profit per cage steadily declines on huge farms. Some survey report provides an overview of the growing rabbit business in China, highlighting its significant economic potential, noteworthy market prospects, and distinctive advantages.

According to Chinese customs, a family that owns several rabbits will never run out of oil, salt, or vinegar; a family that owns 100 rabbits would never run out of garments or pants; and a family that owns a thousand rabbits will never run out of a building. The majority of industrial rabbit production occurs in a few European nations (such as France, Italy, Spain, Hungary, and Belgium), whereas medium-sized and small-scale farms are typical of the remainder of the world, ranging from South America to Asia. Expanding scientific understanding in these kinds of farms is crucial to their growth. Many practical aspects of working on small, medium, and big farms are covered in this chapter. The primary subjects are:

- Breeds and selection
- Rearing future does
- Mating
- Artificial insemination
- Hormonal Treatment and Biostimulation

- Lightning programme
- Doe-litter separation (DLS)
- Suckling mortality
- Teat number
- Seasonal effect

All farms can benefit from discussing most of these topics, with the exception of artificial insemination.

1. Breeds and Selection

Breeds utilized to produce meat fall into different categories:

- Local or native breeds
- Colourful (extensive) breeds,
- Intensive breeds, like Californian or New Zealand,
- Synthetic breeds,
- Hybrids

Some local breeds are mainly employed in nations with hot climates or in particular regions of other countries. Despite having poor reproductive success, they are extremely resilient to keeping conditions (hot climate, low nutritional level, local feedstuffs) and well suited to the local environment. Breed selection does not have a major impact on home meat output or backyard rabbit husbandry. These days, resistance and affordability can be the most crucial attributes.

Many nations raise coloured breeds of rabbit to generate meat for their own consumption or for specific markets. In Europe, some coloured varieties are utilised in alternative or organic farming methods. Their output falls between that of intensive and native breeds. Due of their adequate capacity for reproduction even in harsh environments, tiny farms frequently house them.

Because of their excellent reproductive and productive performance, intensive breeds are appropriate for medium-sized and larger farms. Their consistent litter size and strong maternal competence make them useful. Leading commercial breeds bred worldwide are Californian and New Zealand White.

In order to bring better reproductive and productive traits and adapt to various environmental conditions or market preferences in a given country, two or more intense breeds were crossed to create synthetic breeds. The most well-known artificial breed, the Californian, is descended from crosses between New Zealand White and Himalayan rabbits. In the nation where they are bred, the majority of synthetic breeds enjoy popularity.

The simplest and fastest method is to cross current stock with closely related breeds. Keep in mind that selecting foreign breeds requires caution. The finest options for enhancing reproductive or productive performances could be medium-sized breeds like Californian, New Zealand White, or others. Breeding is another well-known technique. Numerous studies have been conducted to demonstrate the heterosis of crossbred populations in reproductive parameters (such as litter size). Rotational crossing is another method of crossbreeding whereby crossbred females and purebred males are mated. Every purebred sire breed is rotated, or alternated, amongst the following generations:

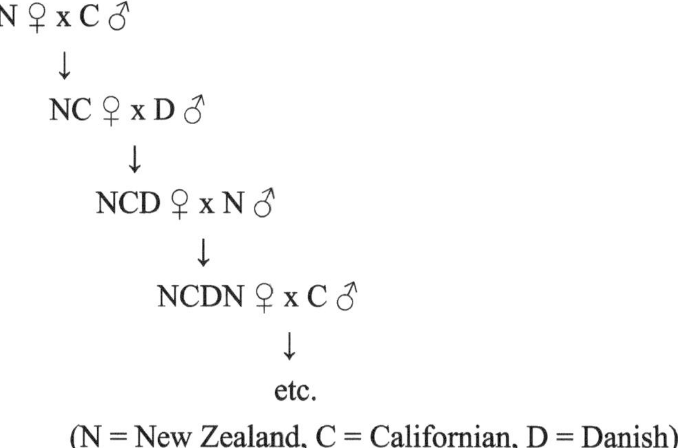

(N = New Zealand, C = Californian, D = Danish)

Selection criteria, methodology, and direct and linked genetic responses in mother lines were compiled by the researchers. Generally, the BLUP process is used as the basis for selection, and the litter size at birth or weaning is used as the selection criterion. Over thirty generations have passed since certain lines were chosen. Because the litter size has a very low heritability (sometimes less than 0.1), there are also very little selection responses per generation (0.03 to 0.18). Certain selection programmes incorporate other factors, such as longevity, genetic variability, or mortality. The primary goal of choosing maternal lines is to create hybrid programme crossbred does (parents). There may be a significant (maternal) heterosis effect (15% and 16% in litter size alive and weaned). Hyplus, Hycole, Hyla, and Zika are the most well-known hybrids; they are composed of two maternal lines and one terminal line.

2. Rearing future does

During the breeding season, the factors that impact reproductive performance are primarily investigated. However, life performance may also be influenced by some conditions prior to first mating. The nutritional state of foetuses, suckling kits, and young females prior to breeding may be one of the most significant consequences. Litter size and the position of the foetuses within the uterine horns are the primary factors that determine birth weight, or the nutritional quality of the foetuses. Numerous studies on humans have shown that birth weight affects longevity and health in the long run. Even so, there are also a number of rodent studies accessible. These animals' birth weights and adult performance are correlated in the same way, therefore choosing individuals with higher birth weights may improve their ability to reproduce.

The kits' nutritional (milk) supply has a long-term impact as well. Small and big litters of kits were raised by researchers. Compared to does born in litters with eleven kits, those in litters with five kits kindled and raised larger litters. According to findings, the size of the litter in which the kits were born and raised also had an impact on how well they performed as adults. Researchers found that the production of does from large litters (12) was poor. There was a noticeable variation in the amount of milk consumed when the researchers nursed kits with either one or two does. The performance of the nursing kits with two does is also improved as they become older. Improved nutrition during the nursing period may lead to improved adult performance. The does' superior condition (greater body weight) may be the primary cause of this. At the age of first mating, the group breastfed by two does had a larger ratio of fat to muscle tissue evaluated by computer tomography than the group nursed by one.

3. Mating

On small and medium-sized farms, natural mating is common and suitable with less than 100 does. The does need to be placed inside the buck's cage. Because the male "lives" in his territory, mating occurs more quickly because he doesn't have to spend time exploring. Examining the vulva prior to copulation is advised. When a doe's vulva is red or somewhat violet and turgid, she is in heat. Researchers found a correlation between the colour of vulva and specific reproductive characteristics (increased conception rate and litter size). The LH and FSH peaks are seen after mating. After mating, ovulation happens 10 to 12 hours later. The sperm travels towards the ampule, the site of fertilisation, at the same moment. The best time for fertilisation is when the sperm and ova cells contact, and they are fertile for 4 and 6 hours, respectively.

Fig. 1: Natural Mating of Rabbit

Some writers advise mating in the morning as well as the afternoon, reasoning that this increases the likelihood of sperm and egg successfully meeting at the appropriate time. Based on our past findings, it appears that mating the does twice with two bucks in quick succession is the optimal method. When does were left in bucks' cages for two hours following their initial mating, the same outcomes were obtained. The rate of service acceptance decreases after mating. The percentage of failed second matings was 0, 5, 7, 16, 19, and 39% when the time interval between two matings was 0, 2, 4, 6, 8, or 24 hours. There were no kits created by the second buck in the previous group. Between 20 and 24 hours following coitus, researchers noted a significant drop in sexual activity.

4. Artificial Insemination (AI)

In Europe, large farms frequently employ artificial insemination, or AI. There are five steps to it.

- Gathering the bucks' semen,
- Looking at the semen under a microscope and visually,
- Diluting semen ten to twenty times if it is of high quality,
- Using a pipette to disperse does,
- Infusing GnRH analogue into the muscle of the hind leg during the insemination process.

Spreading hundreds or thousands of copies in a single day is one of AI's primary benefits. Depending on how the farm is run, the stock may be separated into multiple groups or halved on the day of insemination for all the does. When insemination, parturition, weaning, and selling, among other processes, are all well-organized, AI also makes it favourable to employ cycled production, which facilitates smoother work schedule planning. The entire (empty) rabbitry is cleaned and sanitised on a vast farm. All of the nest boxes are ready, and only pregnant rabbits are kept in cages. Strict inspection (nest box regulating) during kindling helps resolve issues like kindling on wire nett or dispersing of baby rabbits. All patients can simply receive treatment during this crucial time by, for example, adding a combination of water-soluble vitamins to their drinking water.

Eleven to twelve days following artificial insemination (AI), all does are inseminated on the same day, and pregnancies can be palpated. On the same day that they are all weaned, the does and kits are relocated into a different rabbit housing. Medicated feed can be used to prevent coccidiosis and certain gastrointestinal disorders after weaning. All of the growing rabbits are sent to the slaughterhouse at the conclusion of the fattening phase. Does are inseminated 11 days after parturition on the majority of farms in Europe (more than 90% of French farms). This is referred to as the 42-d reproduction rhythm: the overall duration of a pregnancy is 42 days, or around 31 days plus 11 days, between kindling and insemination.

This is referred to as the 42-d reproduction rhythm: the overall duration of a pregnancy is 42 days, or around 31 days plus 11 days, between kindling and insemination. The rabbits will always kindle between Monday and Wednesday (31 to 33 days later) if they are inseminated on Friday. They will also be inseminated again on Friday, six weeks later. 35-, 49-, or 56-d rhythms can be used in addition to the 42-d reproduction cycle (AI 11 days postpartum). Does are inseminated 4, 18, or 25 days following parturition in such circumstances. In Europe, the 35-d rhythm is seldom and excessively intense. Kittens are weaned at 28 to 32 days, while growing rabbits are sold at 70 to 74 days in the 42-day cycle. At this age, the typical weight is between 2.4 and 2.5 kilogrammes. Before the next batch of rabbits is housed, the building can be emptied and the cages cleaned and disinfected in 42+42 days. In order to clean and move new stock, there is not enough time when the slaughter age gets close to 80 or 84 days. A 49-d reproduction rhythm is advised in this situation. At 35 to 42 days, the kits can be weaned, and at 84 to 86 days, the growing rabbits can be taken to the slaughterhouse. When the 42- and 56-d reproduction rhythms were compared, the researchers discovered no appreciable variations between the two groups. In a recent study, biostimulation—altering the nursing approach three days prior to artificial intelligence—was used in a 42-d rhythm group. When a 56-d rhythm was chosen, the kits were weaned at 23 days of age, or two days prior to AI. From the perspective of animal welfare, the 56-d rhythm was preferable (body condition and survival of does) because of the longer time until re-insemination (d 11 or 25). However, the 42-d group produced 19–23% more kits/doe/year than the 56-d rhythm, which had a significant financial impact on rabbit farms.

5. Hormonal treatment and biostimulation

AI is used to inseminate lactating does. Their responsiveness determines the conception rate. In non-receptive does, the researcher found a very strong antagonistic relationship between lactation and reproductive functions. It is advised to use hormonal therapy (eCG or pregnant mare serum gonadotrophin, or PMSG) to increase the responsiveness of nursing does.

The benefits of PMSG therapy include:

- An increased rate of doe responsiveness
- An increased rate of conception
- Greater size of litter
- Its effects are more pronounced in weaker body conditions.

Treatment with PMSG may also have certain drawbacks:

- Ineffectual when used with unresponsive people,
- An elevated rate of anti-PMSG antibodies after high-dosage or frequent treatment,
- Reduced chance of conception following repeated use,
- Greater death rates following birth,
- The majority of customers oppose the use of hormones in animal agriculture.

The adverse consequences of treating with PMSG indicate non-alternative hormonal techniques (e.g. biostimulation). Farms may use three different methods:

- Lighting programs,
- Short dam-litter separation,
- Changing nursing methods.

6. Lighting program

As the daylight lengthens in the spring, wild rabbits begin to go into heat. Their reproductive period officially begins at this point. In rabbitries, extending the daily light period can have a comparable impact. The conception rate increases by almost 10% when daily lighting is increased from eight to sixteen hours, approximately eight days prior to AI. Five days before to AI, the researcher's study saw an increase in lighting from 10 to 16 hours per day, but no discernible difference in the rate of conception. The body weight of weaned rabbits has dropped in each of the ensuing tests. It appears that cutting the dark period short could negatively impact does' feed intake and/or milk output, either directly or indirectly.

When comparing the 8 L:7 D:1 L:8 D lighting schedule with the continuous 16-hour light (16 h light:8 h dark - 16 L:8 D) period, the additional one hour of lighting had no positive impact on does produced. An analysis of the 8 L:4 D:8 L:4 D interrupted illumination programme revealed that while it disrupted the nursing behaviour of does, it had no discernible influence on any reproductive features. Analysing the lighting schedule of 12 L:6 D produced results that were similar.

7. Doe-litter separation (DLS)

Soon after they are weaned, the majority of does mate. Prolactin's suppression of the ovulatory cycle lessens after milk production stops. Does their kit undergo routine nursing once a day? Receptivity increases and hormone changes akin to those seen during weaning can be anticipated if there is a 24-hour gap between two breastfeeding bouts. According to data compiled by the researchers, a 48-hour DLS improves conception rates more than a 24- to 36-hour DLS. If DLS lasted for 24, 36, or 48 hours, the kindling rate increased dramatically (2 out of 5, 2 out of 2, and 5 out of 6 trials by 11, 13 and 16 to 24%, respectively) in the maximum cases utilising free nursing. Nonetheless, in the instance of regulated nursing, the impact of DLS on the rate of conception was not noteworthy. The individual weight of kits at weaning reduces because one suckling episode is lost, even if DLS has no influence on litter size. While DLS has a negligible impact on controlled nursing groups, its efficacy in the free nursing system has been demonstrated. It is possible to attain comparable outcomes by altering the nursing approach. The does get enthusiastic about midnight (the previous nursing time) when free nursing is initially provided and then switched to controlled nursing (nest box opened only in the morning for roughly 20 min). Changes in the nursing method two or three days prior to AI result in 15–27% higher rates of receptivity and conception, and occasionally even larger litter sizes.

Before AI, the most effective methods for 3-d control nursing were either separating the does using a metal plate or removing the litter with the nest-tray 5 m away from the doe (allowing visual contact or none at all) or utilising wire-mesh separation. All three of these approaches are simple to implement on farms and are good substitutes for hormonal (PMSG) therapies. Other approaches were also looked into, but they were either less useful (moving cages, gathering does) or ineffectual (buck effect, feeding programme). The basic performance of does determines the efficacy of biostimulation techniques or hormonal treatment. All of the techniques worked well in trials where the control group's conception rate was between 50 and 60 percent. The difference between the treatment and control groups was negligible when the conception rate was between 75 and 80%. The age (parity order) of the does affects how effective these strategies are as well. After the first parturition, the disparities between the treated (DLS or PMSG) and control groups were greatest. Because the primiparous does had less fat stored for growth, pregnancy, and breastfeeding, their condition was weak in this instance. The rate of conception may also be low in the absence of treatment. Following the third or fourth litter, there are no longer any notable distinctions between the groups. The weakest does—dead or culled—are removed; the animals that survive are in greater health and have the capacity to eat more feed than the younger females. One benefit of AI is its ability to rapidly disseminate high genetic values. One significant concern when starting a fresh rabbit stock is the possibility of illnesses (such Dermatophytes) being spread by live animals. By utilising solely sperm, the "old stock" infection can be reduced.

8. Suckling mortality

To have potential reproductive capabilities, more rabbits must be born overall. In addition, more animals need to be sold or weaned by farmers. The percentage of newborn bunnies who are still alive at weaning is one measure of a doe's ability to reproduce. Pre-weaning mortality is mostly caused by suckling mortality, total litter loss, and stillborn animals. When considering the impact on overall litter loss and suckling mortality, birth weight, litter size, and doe age are identified as the primary contributing factors. Investigating the causes of suckling death in 167 Spanish rabbit farms, researchers found that 13.4% of suckling deaths and 6.3% of stillbirths occurred. The majority of deaths had no medical cause. In contrast to the overall death rate of 100%, 38% of births were stillborn, 19% were starved, and 17%due to hypothermia.

Animals that are too young, too old, or unwell, as well as weak rabbits, have greater embryonic mortality rates. These rabbits are also not very good at raising offspring. In Europe, PMSG is frequently used for receptivity synchronisation despite the negative consequences of hormone therapy. Some does may kindle predominantly tiny litters with a higher total litter loss if they are in poor condition or have fragile health. Superovulation can also occasionally be observed, resulting in the development of sizable litters with a small percentage of low birthweight kits and a subsequent decline in kit viability.

9. Teat number

Most often, rabbits have eight, nine, or ten teats. By genotype, the distribution of teat numbers differs. Depending on the genotype, the frequency of rabbits with eight teats ranged from 23 to 70%, and that of rabbits with ten teats from 13 to 50%. There is a positive association between the number of doe teats and reproductive capacity. In an earlier study, 5–10% more kits were kindled by rabbits with 10 teats than by those with 8 teats. Because the size of the mammary gland is independent of the number of teats, rabbits with more teats are not always more lactating, but they are better at nursing. When does with 8 or 10 teats were compared, the difference in kit mortality was found to be between 5 and 7%, depending on the stock. The length of a nursing session is 2.5 to 3 minutes. Kits can drink 15 to 20 percent of their body weight in milk during this brief period. Kits are incredibly bad. They release the teat and attempt to catch another one as soon as they sense it is empty. The number of teats and size of the litter affect their prospects here. If the doe has only eight teats, two kits in a litter of ten are waiting for a teat every second. Each kit contains ten teats, allowing them to all drink milk.

10. Seasonal effect

In countries with a continental climate, when sunshine and temperature fluctuate significantly throughout the year, the impact of seasonal variations on reproductive and productive capacities is significant. The season has a significant impact on rabbit reproduction in its native habitat, which halts in late fall and early spring. The shortened daytime hours are the cause of this. The extended daylight hours in the spring increase receptivity and conception rates on small and medium-sized farms. Due of the increased embryonic mortality brought on by the high temperatures, fewer kits are born in the summer. A hot environment reduces feed consumption, which in turn reduces milk production and increases the death rate of nursing infants. Rabbits can only increase heat output in response to rising temperatures by vasodilatation, which increases blood flow and breath frequency. The moulting brought on by the shortened daytime hours may be the main cause of the reduced receptivity in the fall. In certain countries, wintertime lows of as low as -10 or -20°C might occasionally pose issues. Shearing the hair off of animals' backs and sides can assist remove heat, as can placing rabbits on wire mesh floors to lessen the effects of a hot climate.

A somewhat warmer environment is produced in the winter by properly covering (insulated) cages and adding additional straw to the cage. There is very little variation in the daylight effect in structures that have 16 hours of lighting every day of the year. Reducing the severe temperature effect can be achieved by cooling the building and sprinkling water on the floor to increase humidity. Large farms are the primary users of heating and cooling equipment. The efficacy of ventilation as a building climate control method is contingent upon the disparity in relative humidity between the exterior and interior of the structure. Ventilation has a significant cooling effect on buildings. The benefit of forced-air cooling is that it eliminates humidity in the air as well as animal heat. A healthy environment is necessary for improved reproductive and productive outcomes.

8

Common Diseases of Rabbits and Control Measures

Many families who have had the pleasure of raising rabbits consider them to be valuable members of their family. Unfortunately, rabbits are susceptible to a wide range of issues and illnesses, much like other pets. Upper respiratory tract infections (snuffles), internal and external parasites, dental disease, gastrointestinal (GI) stasis, uterine issues (cancer or infections), and pododermatitis (foot sores or sore hocks) are common illnesses that affect pet rabbits. Rabbits are incredibly prolific creatures with a vast potential for procreation. Their ability to convert 20% of the protein in their meal into meat protein is second only to that of broiler chickens, who can convert 23%. Since there is essentially no competition between the broiler poultry sector and rabbit rearing for feed supplies, rabbit rearing is predicted to be a viable option.

Just like any other domesticated animal, rabbits can get sick from several things. Numerous infectious and non-infectious disorders affect them. Reductions in mortality, a rise in fryer production per maternity cage, and lower operating and feeding costs are all necessary for the rabbit industry to succeed. In order to maintain a profitable rabbitry, every attempt should be taken to minimise the loss of fryers, does, and kits. The most frequent reason for litter, doe, and fryer losses is disease. Rabbits cannot be treated like other farm animals when it comes to the usual disease control methods used there. This paper aims to provide a concise overview of prevalent diseases in farmed rabbits and their management, particularly for those who are new to the field.

1. Snuffles in Rabbits

Snuffles is caused by a common bacteria, yet it resembles the common cold in humans. Snuffles in rabbitries can be caused by inadequate ventilation. The upper respiratory tract is harmed when an undesirable ammonia odour accumulates in a rabbitry as a result of insufficient ventilation. Rabbits with the clinical illness known as "snuffles" often have nasal discharge, frequent wiping with the inside of their forepaws, coughing, and weight loss. The culprit behind this is *Pasteurella multocida*. There are various strains with diverse backgrounds. Only snuffles result from strain derived from human visits or handlers. However, a highly pathogenic strain of rabbits can result in fatalities, pneumonia, and pleurisy. Direct contact between rabbits on a farm leads to the spread of disease. Stress and inadequate ventilation are important contributors to the spread. Snuffles are an issue in practically all large-scale cage-based rabbitries as well as the majority of small-scale rabbit colonies that are kept in hutches. Antibiotic-containing medications may be able to temporarily resolve the issue. A long-term solution is achievable through methodical culling and management. Ventilation should never be compromised, no matter what.

There are two kinds of ventilation that are frequently used. A power-operated blower is used to bring outside air into the shed, achieving positive ventilation. Utilising ventilators, negative ventilation removes air from the shed. The approach that is most frequently utilised is negative ventilation. The outside temperature should be taken into consideration while placing air intakes.

Fig. 1: Lesion in respiratory tract

While the shed's eaves allow air to enter during the winter, the air that enters close to the floor during the hot summer months is relatively cold. In rabbits, the same *Pasteurella* can result in severe illnesses such pleuropneumonia. Only after necropsy are these grave affections noted. With the exception of weight reduction, living animals exhibit no noteworthy alterations. It is only by palpation that one may identify the worsening state of their body, not through visual inspection.

2. Cutaneous *Staphylococcosis*

Staphylococcus organisms can occasionally result in fatal wet dermatitis and severe abscesses. Staphylococcal moist dermatitis in newborn kits can result in the litter's complete loss. Therefore, before they are weaned, kits should receive extra care. The afflicted kits seem damp, and there are little abscesses visible everywhere on the body. In large rabbitries with poor litter hygiene, this disease is rather prevalent. *Staphylococcal* pathogens can occasionally impact both does and kits. The affected individual does have mastitis. The *staphylococcus* strain that affects humans is moderate, and abscessation happens sometimes.

On the other hand, the sickness in young kits is more severe and deadly due to the rabbit strain. When a doe has mastitis, the kits usually lose their milk supply, get malnourished, and eventually die. Mastitis in rabbits is commonly referred to as "blue bag." The kits can be prevented from malnutrition and death by early identification. In addition, uterine inflammation and visceral abscessation in the thoracic and abdominal cavities are symptoms of staphylococcosis. On damp skin, the lesions appear as pin-head-sized, superficial white foci. Many pea-sized abscesses under the skin, purulent conjunctivitis, closed eyelids from sticky exudates, and pneumonia are common observations in older rabbits. The discovery of a *staphylococcal* abscess in a newborn litter is detrimental to the survival of big rabbitteries since it is extremely difficult to eradicate. The occurrence can be decreased with proper litter management, nest box hygiene, and prompt culling of afflicted animals. Antibiotics such as tetracycline have been found to be an effective kind of treatment.

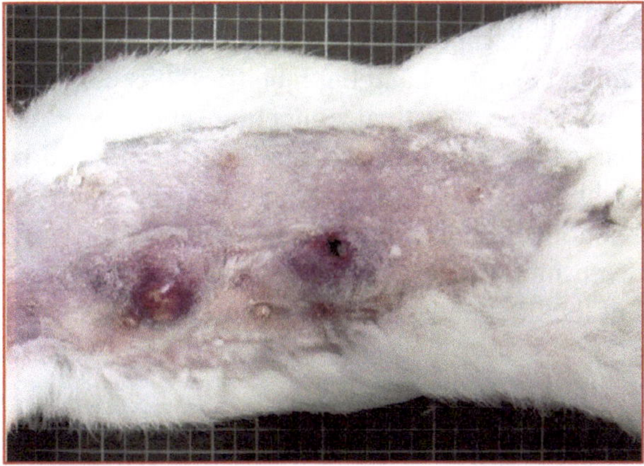

Fig. 2: Lesion on skin

3. Abscesses under the skin

Pasteurella affected rabbits may experience numerous subcutaneous abscesses across their body. In every age group, this is a rather regular occurrence. This illness is commonly observed in both big and small-scale rabbitries. Subcutaneous abscesses can vary in size from the size of a peanut to that of a hen's egg and can occur in various body areas. Abscesses can appear as a single lesion or in groups.

Fig. 3: Typical lesion of abscess

The two primary bacteria that cause abscesses in rabbits are *Staphylococcus aureus* and *Pasteurella*. Strangely, the pus from rabbits is thick and sticky, similar to dental paste. Draining through a tiny hole is therefore challenging. To release the pus, the abscess must be opened widely. Tetracyclines and other systemic antibiotics are helpful in treating rabbit abscesses.

4. *E. coli* Infections

Both weaned and unweaned rabbits can contract *Escherichia coli* infections. Yellowish diarrhoea, inappetance, and mortality are its hallmarks. Infections with *E. coli* have been reported in both small and big rabbitries. It is common for *E.coli* organisms to adhere to the intestinal epithelium and cover the surface of the absorptive villus. As a result, absorption fails due to a reduction of absorptive surface. Kitten rabbits, ages 1 to 14 days, are extremely delicate. Watery diarrhoea, a bloated abdomen, moist and discoloured posterior body parts, loss of appetite, and even mortality are signs of an *E. coli* infection. Severe outbreaks destroy all of the litter and damage the entire nest.

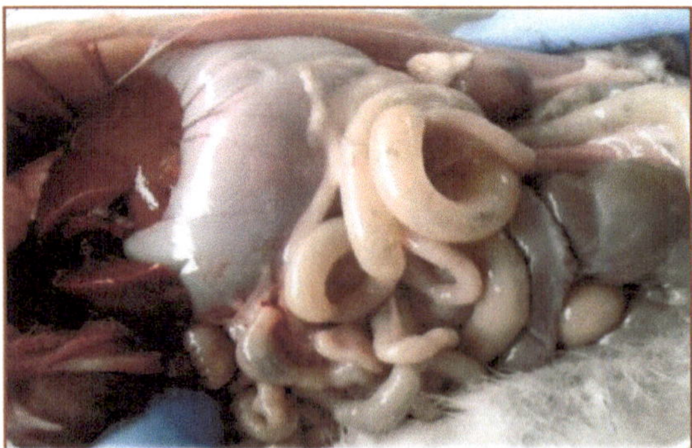

Fig. 4: Infected rabbits show diarrhea and emaciation

5. Wry neck condition in Rabbits

Pasteurella multocida also causes this illness. *Pasteurella* organisms not only cause snuffles, pneumonia, and cutaneous abscessation, but they also damage the middle ears, which results in a characteristic head tilt. The inner ear becomes infected in severe cases, and the middle ear is typically infected. Rabbits with the condition maintain a head tilt towards the affected side. If the affected ears are closely examined, it may be possible to see that dried exudates have formed a scab and are obstructing the ear canal.

Fig. 5: Infected rabbit shows Tilt Head

The mediastinum, peritoneum, and uterus are among the bodily cavities affected by certain highly pathogenic strains of pasteurella.

6. Coccidiosis

Protozoan parasites known as ccidian parasites infect rabbits. They are often used in traditional colony keeping. Coccidiosis frequently causes development retardation in broiler rabbits. Sometimes, serious infections result in paralysis of the rear limbs. The incidence of coccidiosis is almost usually linked to unsanitary living conditions. The incidence of cocciodiosis was significantly decreased by using coccidiostats and keeping the rabbits in cages. In rabbits, there are two different types of coccidiosis. These two conditions are intestine and hepatic cocciodiosis. *Eimeria steidae* is the causative agent of liver cocciodiosis, which can arise in both conventional and commercial rabbitries. Often impacted are young bunnies. However, growth retardation is the more prevalent symptom and death is rare. Liver cocciodiosis is frequently discovered by accident during postmortem testing. Typical necropsy findings include an emaciated carcas, a moderately to severely enlarged liver, and pea-sized abscesses in the liver. Sulphonamides are an excellent treatment for rabbit coccidiosis.

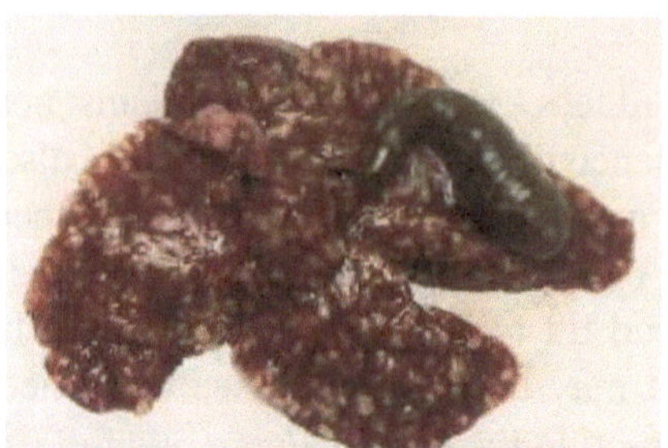

Fig. 6: Lesion of Hepatic Coccidiosis

Eight different species of *Eimeria* are responsible for another type of the disease known as intestinal coccidiosis.Two of these, *Escherichia coli* and *E. intestinalis*, are very virulent and can be fatal. Some species, such as *E. magna, E. irresidua, E. perforans, and E. coecicola*, are less harmful and frequently cause growth retardation. Upon necropsy, the intestinal wall thickening and the presence of pinpoint haemorrhages in the colonic and caecal mucosa are typical. Separating intestinal cocciodiosis from clostridial endotoxemia and *E. Coli* infection is important. Using commercial coccidiostats in feed, treating afflicted animals and isolating them, putting the rabbits in wire mesh floor cages, avoiding straw bedding, and other hygienic measures can all effectively decrease the incidence. It is recommended to administer sulphonamides for seven days in a row in order to avoid coccidiosis, and then to repeat the treatment after a seven-day break.

7. Sore Hock

Alternatively known as ulcerative pododermatistis, this disorder is triggered by a variety of reasons. The rabbit's skin is extremely delicate and silky. As a result, they are more vulnerable to cuts, abrasions, and other ailments. The factors that determine the frequency of sore hock in rabbits are skin thickness, paw hair cushioning, wire mesh quality, and floor litter. Minor cuts have the potential to develop into severe ulcers or even lymphangitis. Abscessation is a typical aftereffect. Symptoms of painful hock include restlessness, hunger loss, weight loss, and discontinuation of breeding activities. Staphylococcus aureus, the causal agent, may also spread to the abdominal cavity and, in more severe cases, result in kidney lesions. To some extent, this is preventable.

There's no doubt that changing frequently and using dry bedding will lessen the occurrence. In a cage floor, only appropriate breeds or crossbreeds should be housed. The prevalence of sore hock has significantly decreased since the advent of plastic grills. Treatment for broiler rabbits is not common; local and systemic antibiotics and sprays are only used on valuable breeder stock.

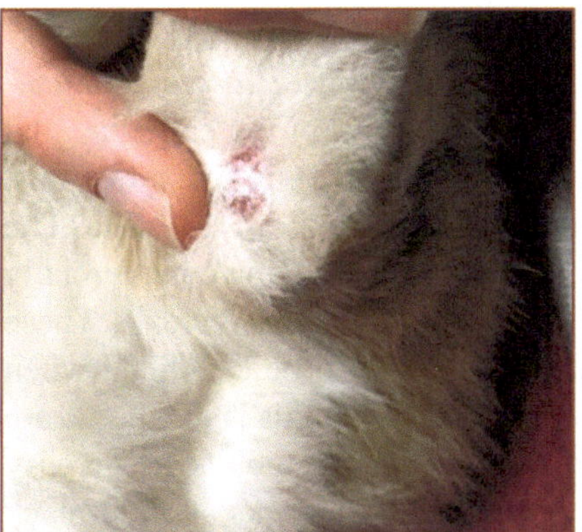

Fig. 7: Lesion of Sore hocks

8. Caecal Impaction

Similar to the rumen in cattle, the posterior portion of the alimentary tract plays a similar role in rabbits. Fermentative digestion is facilitated by the distinct motility pattern and bacterial population of the caecum in particular. Impaction results from the caecum's regular digestion process being hampered for unclear causes. The deposition of thick, viscous mucus and subsequent slowing of intestinal motility are the defining features of this illness. A gradual decrease in motility leads to cecal constipation. Since the ailment had a non-inflammatory aetiology, research on it has come to an end. Therefore, the medical term for this illness is complex mucoidenteropathy with caecal impaction. Two weeks after they are weaned, rabbits have ileal impaction, which is most commonly observed in stall-fed, high-concentrate, low-roughage diet settings. Illness may come on suddenly or gradually, and impaction will almost certainly progress.

The affected animal keeps itself apart from its cagemates. They quit consuming roughages and pellets, but they keep drinking more water. Teeth grinding is often observed. Usually, a mass that resembles a cable can be felt when palpating the right ventral abdomen. Fluid coming out of the nose is a rather common observation in affected rabbits just before they pass away. The mortality rate is between 60% and 100%. Lesions found after death are exclusive to the caecum. The symptoms of caecal impaction include a mass of mucus in the caecum and a dried-up or watery substance. Periodically, abnormalities in the lung's apical lobe may indicate foreign body pneumonia. Most of the time, treatment is unsuccessful. Dexamethasone and metronidazole/analgin are used with varying degrees of success. The dose of tetracycline used for prevention is 250 mg per litre of drinking water.

9. External Parasites

Mange, or mite infestation, is prevalent in backyard rabbits. *Psoroptescuniculi* is a mite that is often recognised. Greyish-white crusts in the ear canal and ear lobe, head shaking, itching ears, and claw scratching are the symptoms. Applying a mixture of ectoparasiticide and paraffin can solve the issue. The problem can also be cured by removing the crusts and applying commercially available ointments (containing both antibiotics and ectoptasiticidal). These days, subcutaneous ivermectin injections are reported to be beneficial when repeated after six days. The parasitovorax and chyletiellaThe rabbits are afflicted with fur mites called Listrophorus gibbus. White rabbits are an easy way to identify this infestation. appears on the bunnies' backs as little black spots. Itching is often accompanied by symptoms. It works well to apply topical treatment or to dip using commercial dipping solutions.

In rabbits, lice and fleas are uncommon. Myxomatosis is a highly contagious viral disease that previously decimated significant populations of wild rabbits in Australia and New Zealand. The European rabbit flea, Spilopsyllus cuniculi, is common and is a powerful mechanical transmitter of the disease in rabbits. In Australia, the myxomatosis outbreak was purposefully dispersed to lower the number of wild rabbits, which were becoming a nuisance to farmers.

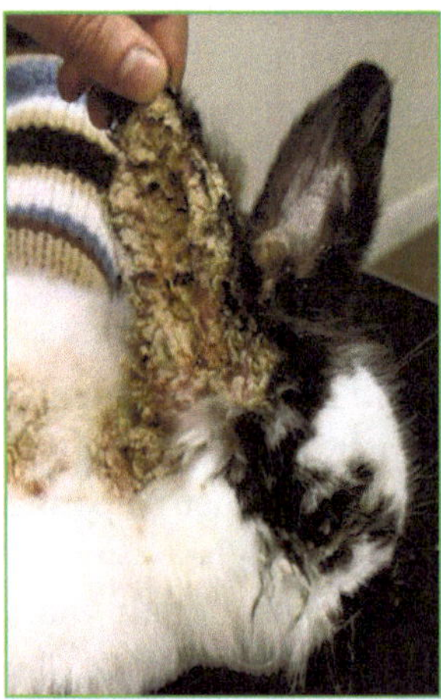

Fig. 8: Parasitic infestations in ear

10. Hairballs / Wool Blocks

Rabbits are naturally extremely clean creatures. They groom themselves on a regular basis. They may ingest fur during grooming, which can result in trichobezoars. In a physiological sense, they do remove their fur from the ventral belly, the area around the mammary glands, the inner thigh, and the dewlap as they are building their nest in the later stages of pregnancy. The nest is lined with the fur that is removed. Additionally, there is a chance that the fur will be swallowed during this process. An extremely low fibre diet puts rabbits at risk of consuming fur. Hairball development and intestinal blockage are the results of all these conditions. A thorough abdominal palpation can be used to diagnose this illness. A serious obstruction could endanger one's life.

The wool block can be removed by feeding some loosening agents, such as mineral oil (20 ml), and gently massage. The hair balls can also be broken up by immersing 10 ml of fresh pine apple juice two or three times a day. The hairball is loosened by the enzyme bromelin, which is found in pine apple juice, and then eliminated naturally through faeces.Similarly, papain found in commercial pharmaceutical goods or papaya juice can likewise be put to use. The prevention of hairballs in rabbits can be achieved through the provision of sufficient fibre, alleviation of boredom, and alteration of the light cycle through lighting pauses.

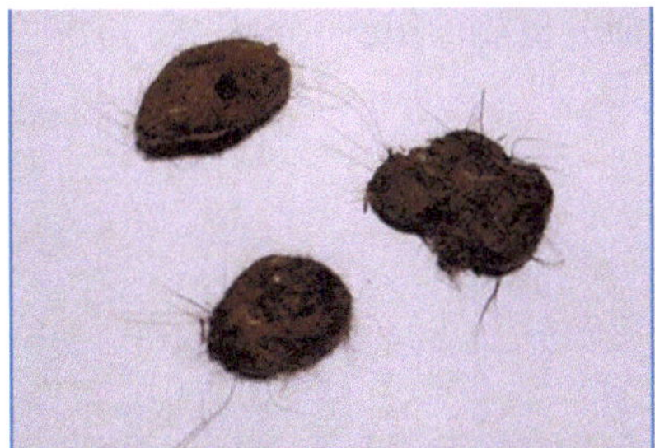

Fig. 9: Hairballs ingested by Rabbit

11. Myxomatosis

This is a contagious viral illness that causes skin tumours in rabbits. There could be minor lesions occur in both domestic and wild rabbits. Severe manifestations could be fatal. Ticks, fleas, mites, and flies are among the ectoparasites that spread the myxoma virus to rabbits housed in colonies. Iatrogenic transmission, or the sharing of needles and other equipment between healthy and sick rabbits, is a frequent occurrence in big rabbitries. As mechanical vectors, fleas and mosquitoes are involved. Based on the seasonal breeding of vectors, this disease manifests itself in different seasons. Oedema of the head, eyelids, genitalia, blepharo-conjunctivitis with purulent discharge obstructing the movement of the lids, and blindness are the typical symptoms of acute instances of myxomatosis.

Pseudo-tumors in the ears, nose, and paws are frequently observed in cases of persistent myxomatosis infection. They might deplete and regenerate on their own. These pseudo-tumors spontaneously retreat, leaving behind a scab that may eventually fall off. In several nations, vaccination is a practice. In a vaccine outbreak, iatrogenic transmission is also frequently observed. In an outbreak, immunisation is not recommended, however the severity of myxomatosis may decrease. To stop the spread, every maternity cage needs to have its needle changed.

12. Toxoplasmosis

A protozoan parasite that starts in cats can spread to rabbits, humans, and other domestic animals. There is a higher seroprevalence of Toxoplasma gondii than clinical disease. Strict preventive measures must be implemented for toxoplasmosis due to its zoonotic potential. The main source of infection is cat faeces contaminated feed. Fever, listlessness, and death within a few days are noted in cases of toxoplasmosis. The rabbits that are pregnant or nursing have the highest morbidity. Organ congestion, inflammatory lung lesions, and a severely enlarged liver (ten times its usual size) are all visible during the necropsy. Potentiated sulphonamides work well in this situation.

Fig. 10: *Rabbit infected with Toxoplasma gondii*

13. Ring Worm

Fungi called Microsporum and Trichophyton are frequently responsible for ring worm lesions. Direct touch and fomites shared by infected and healthy rabbits are the two ways they spread. They result in skin that is covered in round, hairless patches. Scales eventually appear on the lesion. These agents are fungal and zoonotic. Diagnosis is by using a wood lamp examination of the lesion or a microscopic analysis of the hairs that were plucked. Miconazole and other topical keratolytic drugs work well as antifungal medications. When dietary and management deficiencies are addressed, lesions typically resolve on their own, even in the absence of medical intervention.

14. Encephalitozoonosis

A protozoan parasite called *Encephalitozooncuniculi* causes neurological illness in rabbits. The symptoms include convulsion, torticollis, paralysis, incontinence during urinating, and burning in the perineal region. Kidney lesions, tiny, depressed patches of necrosis, and scars on the renal surface are frequently found during necropsy. Encephalitozoonosis frequently results in latent infections.

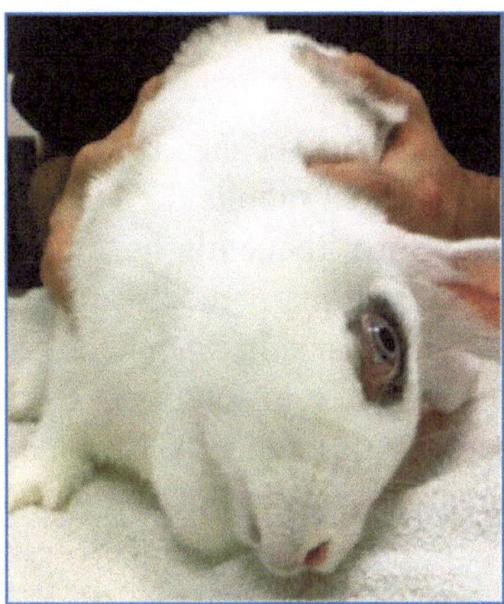

Fig. 11: Rabbit infected with *Encephalitozooncuniculi*

15. Rabbit Syphilis

The cause of syphilis in rabbits is *Treponema cuniculi*. Infections with spirochetes are uncommon in industrial rabbitries. This agent can only infect rabbits; humans cannot contract this sickness. Syphilis symptoms include erythema (redness), edoema (swelling), vascular lesions on the limbs and external genital organs, and scabs from ruptures. Autoinoculation may cause lesions to eventually spread to the mouth and muzzle. Ulcerations caused by self-mutilation are excruciating. To identify Treponema, a skin scraping must be examined under a dark field microscope. In rabbits, this is one of the illnesses that responds quite well to treatment. When taking tetracycline with chloramphenicol, a full recovery is anticipated.

16. Moist Dermatitis

This ailment is brought on by *Pseudomonas aerogenosa*. Not seen all that often. Typically, eczematous lesions cause a change in fur colour that is greenish in colour. This problem is resolved by topical and systemic treatment using standard antibiotics.

17. Tyzzer's Diseases

The bacterium that causes this is *Clostridium piliforme*. The symptoms of Tyzzer's disease are diarrhoea, condition loss, and mortality. The manifest-alike cases of colibacillosis, cocciodiosis, salmonellosis, and clostridial septicaemia complicate clinical identification. A faecal smear examination may reveal cocci grouped in piliform or filamentous stems, indicating damage to the lower alimentary tract. On the surface of the liver, necropsy revealed salt-grain-sized abscesses.

18. Red Urine

Clinically healthy rabbits typically have turbid, murky urine. This is brought on by the urine's high calcium carbonate content. Caged rabbits do defecate in the same spot on a regular basis, and the cage floor frequently has layers of chalky calcium carbonate. Rabbits occasionally urinate reddish-tinged urine. It's not clinically significant, but farmers are interested in it nonetheless. Plant metabolites and pigments are typically the cause of crimson urine. Even among cagemates who share the same food, there are differences. There may be a theoretical relationship between feeding greens like spinach, carrots, and radishes and red urine.

There is no need for therapy because this ailment gets well on its own. Rarely, red colouring of urine can also be caused by urinary tract infections, uroliths, and uterine infections. A definitive diagnosis must be made using certain diagnostic techniques.

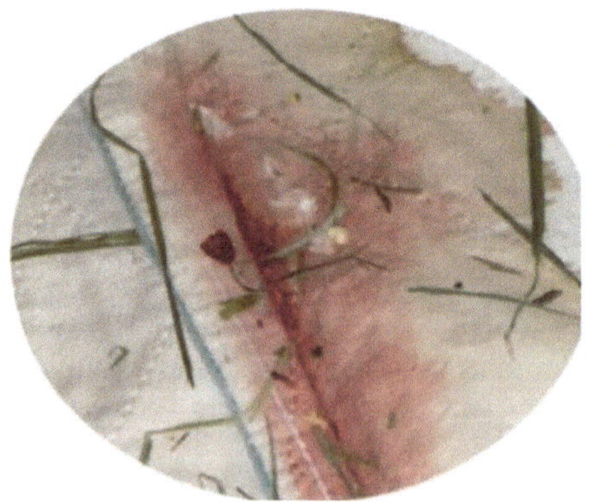

Fig. 12: Red urine

Table 1: Common rabbit illnesses: how to prevent and manage them in a nut sell

S. No.	Name of the disease	Important Symptoms	Prevention and control measures
1.	Hairfball occlusion	Wool builds up in the stomach and prevents food from passing through normally.	(i) Treat the animal with mineral oil or neopeptin or fresh pineapple juice.
2.	Body mange (Ear canker)	Severe pain, itchiness, skin scaling, fur loss, and weakening of the animal.	(i) Application of ascabiol lotion (ii) Ivermectin injection 0.02 ml/kg body wt (S/C) (iii) Strict hygienic measures
3.	Enterits complex	Subnormal body temperature, hunched over, tangled hair, viscous mucus, dry stools, and edoema.	(i) No successful treatment (ii) Tetracycline in feed @ 100 gm/tonne feed (iii) Adequate amount of fibre should be given.
4.	Pasteurellosis (Snuffles)	Discharges from the nose, paw rubbing, appetite loss, and elevated body temperature.	(i) 400000 IU of pencillin and 0.5 g streptomycin given I/M
5.	Coccidiosis	Dehydration, appetite loss, and diarrhoea. may result in mortality in a day or three.	(i) Coccidiostats like Sulphaquinoxaline + Sulphamerazine @ 0.02 to 0.10% in drinking water (ii) Strict Hygienic measures
6.	Shorehock	Inflamed areas or sores on the under surface of hind paws, weakness and dehydration.	(i) Sores should opened and antiseptic dressing should be applied. Provide soft bedding during treatment period
7.	Hind quarter paralysis	Sudden jerk, excitement, paralysis, loss of control over urination and defication.	(i) Affected animals should be slaughtered.
8.	Wryneck	Weakness, inability to eat, and persistent neck twisting to one side.	Affected animals should be slaughtered

19. Indirect Antibiotic Toxicity

Penicillins should not be used in rabbits. Amoxicillin, ampicillin, vancomycin, tylosin, penicillins, and lincomycocin can all indirectly poison rabbits. When these antibiotics are used orally, they disrupt the regular ecosystem of the caecum, changing the process of fermentative digestion. When there is no typical fauna, microbes resembling Clostridium piliforme proliferate quickly and an endotoxaemia-like crisis takes hold. It takes a few days to two weeks for this pathogenesis. Rabbit deaths are typically caused by indirect antibiotic poisoning. Tetracycline, aminoglycosides, fluoroquinolones, sulphonamides, and chloramphenicol, on the other hand, are relatively less toxic to rabbits.

Preventing general illnesses in rabbits

Preventing general diseasesThe three essential components of a successful, return-focused rabbitry are observation, ventilation, and cleaning. Inadequate care can lead to diseases in rabbits. Nearly every disease that affects rabbits is brought on by poor management, either directly or indirectly. Every Rabbitry has three essential components that must be tightly adhered to. They are observation, ventilation, and cleaning.

Cleanliness

It is the main strategy for keeping illnesses from striking rabbits. It is crucial to dispose of garbage, litter, animal faeces, dead animals, and kindling remnants properly and within a set time frame. They may draw predators in addition to being a health risk.n Rabbits

Ventilation

The excrement of rabbits smells strongly. There are bare calcium carbonate deposits in their urine. The output of ammonia is high. When there is inadequate ventilation, ammonia builds up within the home, which has a negative impact on health and ultimately productivity. One of the main causes of snuffles, a common pasteurella infection in rabbits that can seriously endanger rabbitry, is poor ventilation.

Observation

Rabbits are silent creatures that subtly display clinical disease. Regular visits and observation in rabbit houses enable us to identify illnesses early and provide timely medical care. Myxomatosis can be stopped from spreading if exogenous parasites are identified early. The affected animals ought to be kept apart. Culling and disposal are recommended in order to stop the spread of pasteurellosis, staphylococcosis, and many other infectious diseases.

9

Marketing and Profits

Rabbit-based cuisine has been made and enjoyed on a regular basis since the first early civilisations in the Mediterranean region. Rabbit is an easy-care plant that grows well in backyards and on farms. There are notable differences in rabbit production between nations. In 2020, Asia produced 70.5% of the world's rabbit meat, with Europe coming in second. Based on estimated figures obtained from FAOSTAT, there was a 24.1% decline in the global production of rabbit meat between 2010 and 2020. Nonetheless, various patterns emerged throughout several continents. Europe saw a significant decrease in the output of rabbit meat (41.2%), whilst Africa saw a rise (+23.5%). The top two producers are the Democratic People's Republic of Korea and China.

Data obtained from FAOSTAT shows that while rabbit meat output in the Democratic People's Republic of Korea increased from 133,900 to 142,769 tonnes (+6.6%), it declined gradually in China from 690,000 tonnes in 2010 to 456,552 tonnes in 2020 (33.8%). There are many creative substitutes for rabbit meat on the market, such as sausages made from rabbit meat that are smoked, canned, frozen, cured, sauce-picked, dried, and roasted. Meats from rabbits are high in protein, low in fat, high in unsaturated fatty acids, and low in sodium and cholesterol. Compared to several common varieties of red meat, rabbit meat has considerably greater energy values (899 kJ/100 g in the forelegs and 603 kJ/100 g in the loin). This is due to its high protein content, which accounts for 80% of the diet's overall calories for humans. As a result, elderly individuals, teenagers, and pregnant women should all consume rabbit meat. The purpose of this chapter is to give a general review of marketing and consumer preferences. The examination of important aspects of rabbit farming and consumption can have practical ramifications for farmers, business, marketers, and legislators. These can serve as the foundation for the adoption of tactics meant to advance the rabbit meat sector.

Rabbit Meat Production

Sustainability in Rabbit Meat Production

When considering sustainable practices in the social, economic, and environmental spheres, raising rabbits can be an excellent substitute. The majority of rabbits raised for meat are raised in rural areas, where it provides farmers with economic options and helps to maintain population levels in marginal areas. Therefore, rabbit farming helps to keep rural economic activities alive (such as the manufacture of meat, leather, feed mills, etc.) and halts the trend of rural regions being abandoned. There are also significant aspects of rabbit breeding that are related to environmental sustainability. A method of producing meat must be efficient in order to be sustainable in the long run. Efficiency is directly tied to an animal's capacity to turn feed into meat. Rabbits may, therefore, convert 20% of the protein in their diet into muscle mass; this conversion rate is comparable to that of fowl (22%).

In comparison to other livestock like beef, where it fluctuates between 8% and 12%, and swine, where it varies between 16% and 18%, the efficiency of protein transformation into meat is higher in rabbits. Therefore, one of the most important traits for forecasting the financial success and environmental sustainability of the farming system is the animal's capacity to turn feed into meat. Furthermore, rabbits are able to adapt to a wide range of environments and can consume food scraps and agricultural byproducts. A crucial prerequisite for farms, feed mills, and slaughterhouses is traceability. Depending on the nation, farmers, feed mixers, and slaughterhouses collaborate to varying degrees along the supply chain. They don't really depend on one another. In other instances, a single proprietor oversees the farm, the slaughterhouse, and the feed mixer. To win back customers' trust in beef products, many firms are updating their voluntary traceability systems and label indicators. Their goal is to provide and ensure product quality and safety by keeping an eye on the whole food supply chain, from farms to retail locations. The identity of rabbit carcasses from a farm is maintained by labelling and documenting. However, without organisational support and technical help., individual farms and slaughterhouses may find it difficult to set up a traceability system on their own.

Factors Affecting Rabbit Meat Quality

The meat from rabbits is often of a consistent quality. Despite being selected for delayed growth, rabbits have a higher muscle-to-bone ratio and a better-shaped carcas. While oxidative parameters and free fatty acids are not influenced by genetic makeup, animals selected for fast growth rates have higher fat content and lipolytic activity in the flesh of their hind legs. The growth rate and feed efficiency decreased two months prior to slaughter. Therefore, it seems that the most economical way to produce rabbit meat is to wean the animals at one month of age and kill them at two months of age.

Season and temperature are the main issues. Feed consumption is reduced as the ambient temperature rises above the thermoneutrality value. As a result, it causes the rabbit to grow less and weigh less. Because there is less skin, empty gut, and offal at commercial slaughter age, a higher ambient temperature may, under some circumstances, result in a higher slaughter yield. In a similar vein, temperature below the thermoneutrality threshold also affects growth rate because of the current thermostatic intake control system and the additional energy required for thermoregulation. The reduction of seasonal effects on growth and yield can be achieved by controlling the environmental temperature within the range of thermoneutrality. The main ingredients of meat, excluding water, are lipids and proteins. Rabbit meat has a high biological value, is high in protein, and contains large amounts of essential amino acids. Additionally, meat is a good source of easily absorbed micronutrients, such as minerals and vitamins. Furthermore, rabbit meat has a low purine level and is free of uric acid. Based on the portion of the carcas and the numerous productive factors, especially feeding factors, which have a strong impact on the chemical components of rabbit meat, especially on its lipid composition, the information currently available on the chemical components of rabbit meat is highly variable, especially in terms of fat content. The effects of pressure cooking, roasting, frying, steaming, sous-vide cooking, boiling, and microwaving on the nutritional value and palatability of rabbit meat were investigated. Depending on the processing method, cooked rabbit meat's physicochemical characteristics varied significantly. Deep-fried rabbit flesh was fairly hard, chewable, and had a bright golden colour. Roasted rabbit meat is harder and easier to digest. Sous-vide rabbit meat showed reduced flexibility and fewer cooking losses.

The materials used, lighting, and stocking density are the main hazards associated with the housing system. Lines with greater development rates and muscle mass have been produced by genetic selection in the rabbit. Breeds, on the other hand, tend to be less strong and behave more nervously, which puts other rabbits in danger of being frightened or hurt. Poor quality feed and water, as well as imbalanced diets, are further considerations. Water and feeders should be easily accessible to the rabbit in order to provide high-quality rabbit meat. Another factor to think about is biosecurity, which is the prevention of illnesses and infections on farms. Poor welfare circumstances, insufficient feed and water, and unhealthy conditions for the does should all be avoided in reproduction management. Avoiding unfavourable ambient circumstances is also advisable. Rabbit meat quality is also influenced by the housing arrangement, which can involve pens, cages, or outdoor spaces. Depending on their age and size, different varieties of rabbits may experience specific issues with each system. Furthermore, concerns about animal welfare might have an impact on the quality of meat. These might be behavioural as a result of circumstances that make it difficult to relax, elicit anxiety, or restrict movement because of limited room. The following health-related variables may have an impact on the quality of rabbit meat: hunger, thirst, injuries, reproductive issues, and heat/cold stress.

Rabbit Meat Supply Chain and Marketing

Over the years, there have been significant organisational and logistical changes to the rabbit supply chain. Before the industrial revolution, farmers would either sell their rabbits straight to the market or to butchers. There were numerous urban systems for raising and killing animals in rural regions, like Spain. Following then, their yields increased dramatically to satisfy both consumer demand and the growing number of cities. Farms have to focus on certain goods instead of becoming their identity, such selling rabbit carcasses to restaurants, small shops, and butcher shops in towns and cities. Another major turning point in European agriculture came with the introduction of specific breeds and industrialised feed following World War II. This was particularly true in Europe, where eating rabbit was already common. Consequently, the number of rabbit-producing companies decreased significantly, but their size increased continuously. A variety of metrics are frequently employed to ascertain an item's commercial market worth. One of these is the commodity's market share, which is the proportion of money or physical units it represents in the industry. While broiler prices typically vary from 2.50 to 5.95 euros per kg, the average market price of rabbit meat in European butcheries and poulterers is between 5 and 10 euros per kg. Consequently, there is a natural fall in the consumption of rabbit meat, especially in lower-class households. at the meantime, the most prevalent issue is the scarcity of rabbit meat at butcher shops, meaning that city dwellers who are fond of rabbit meat may only find it in a few locations. The main obstacle to producing rabbits sensibly is the increased price of commercial feed, which is 20–30% more than for pigs and double that of broilers. This is partially due to the fact that feed is made up of foods like corn, soybean meal, dried alfalfa, and other feed ingredients.

Usually, 30 to 40% of a rabbit's diet consists of dehydrated alfalfa and 10 to 20% of it consists of soybean meal. Contrarily, consumer opinions are employed to evaluate the food industry. In particular, two principles help us assess it. The "image" and "placement" are these. Customers use important characteristics to define a product or brand's image. Put another way, brand perceptions and characteristics in the marketplace are reflected in placement. After a product is dissected into its component variables, attributes are utilised to define or illustrate the product. When assessing rabbit meat, consumers primarily look at taste, healthfulness, cost, provenance, and technique of production. The definition of a brand's or company's image is how consumers view the brand or company, including its products. It can rely on several factors, such as communication, brand identity, and the calibre of the product. Regarding producers of rabbit meat, factors such as feed quality and circumstances for animal welfare may play a significant role in defining the company's reputation. When a comparable feature is connected to multiple brands or products, the positioning of qualities and brands helps analyse their relationship and can identify brand competition. In a similar vein, the presence of a characteristic in the absence of a surrounding brand indicates fresh marketing chances.

Consumer Preference and Demand for Rabbit Meat

The amount and kind of meat consumed are being impacted by changes in socio-demographics, incomes, eating habits, and costs as well as growing consumer concern over environmental, health, and animal welfare issues. The two main motivators among them are population increase and income. There is a movement towards more expensive sources of protein due to financial availability. According to projections, there will be an 11% increase in population by 2030, which will lead to a 14% rise in meat consumption. Compared to wealthy countries (+0.4% in Europe and +9% in North America), emerging countries (+18% in Asia, +12% in Latin America, and +30% in Africa) are expected to have higher growth in meat consumption. More information about meat demand can be found in the analysis of annual per capita consumption. When only cattle, sheep, pork, and poultry were taken into account, the average amount of meat consumed globally per person in 2020 was 34.44 kg. With 14.78 kg and 11.42 kg of meat consumed per person, respectively, poultry and pig have the largest per capita consumption, followed by beef (6.41 kg) and sheep (1.82 kg). The apparent consumption per capita can provide further information regarding the consumption of rabbit meat. In comparison to poultry, pork, and beef meat, the projected annual per capita consumption of rabbit meat worldwide is 0.19 kg. However, each country has a different level of consumption per person. The Democratic People's Republic of Korea (6.81 kg), the Czech Republic (3.74 kg), Spain (1.09 kg), and Italy (0.91 kg) have the highest weights. Thirty-four to thirty-five percent of consumers do not eat rabbits. The average frequency of consumption of rabbit meat (29–79% of survey respondents) is one to three times a year, which is less than that of other meat varieties.

The demand for and frequency of consumption of rabbit meat are influenced by a number of factors, including the socio-demographic traits and nationality of customers. Men are more likely than women to eat rabbit meat, according to previous studies. Higher income levels and older age groups were found to consume more rabbit meat. Rather, a larger consumption of rabbit meat was linked to a lower level of education. It was discovered that one of the main factors influencing national variations was culinary tradition. Consumption was higher in Mediterranean nations—where rabbit farming has ancient roots—than in nations like North America, where there are no well-established culinary customs surrounding this meat.

Consumption of Rabbit Meat and Consumer Attitude

The attitude of consumers, which is characterised as their favourable or negative assessment of a behaviour, is a strong predictor of their consumption of rabbit meat. Consumer views and opinions regarding the eating of rabbit meat influence attitudes. Examining attitudes and ideas towards rabbit meat is essential to comprehending consumption patterns across many nations. Positive attitudes on rabbit meat influence attitudes, which in turn influence how much of the product is consumed. The most significant attitudes and ideas around rabbit meat are related to aspects of the product, including sensory qualities, cost, convenience, healthfulness, and ethics. Studies on customer preferences on the sensory qualities of rabbit meat produced contradictory findings. Those who ate rabbit meat reported that it was soft and had a nice, subtle flavour. Conversely, those who steered clear of rabbit meat thought it tasted bad. In addition, customers—particularly the younger ones—thought rabbit flesh looked repulsive, especially when it was offered for sale as a complete carcas with the head. A few research examined opinions on the nutritional content of rabbit meat in relation to perceptions of its healthfulness. A subset of customers believed that rabbit meat was healthier than fish and other meats. Customers' opinions of rabbit meat's reduced fat and cholesterol content were very favourable. The majority of customers, however, were unaware of the product's many advantageous dietetic features. Educating customers on the health benefits of rabbit meat's nutrients will change their perception of the product. Convenience-focused products are becoming more popular as a result of recent social and economic developments. A product's convenience features are all those that enable users to save time and effort when engaging in any food-related activity. There are plenty of beef and chicken alternatives to meat that are ready to cook and eat on the market, but there isn't a large selection of rabbit meat products with convenience characteristics. Because cooking rabbit meat involves following traditional local procedures that call for a lot of time and skill, the meat is typically sold as a complete carcass. The abundance of bones in rabbit meat is another factor that affects how easy it is to consume.

Customers, particularly the younger ones, believe that eating rabbit meat can be somewhat challenging due to the small size and fragility of the bones. Customers are becoming more aware of animal husbandry practices in recent years due to ethical and health concerns. The numerous instances of food scares linked to animal diseases and the growing concern for environmental and animal welfare issues are the reasons for this increased interest. Systems involving animal incarceration and intensive livestock production are not well-liked by consumers. Regarding this, the goal of organic certification is to ensure that greater standards of environmental and animal welfare are met. A significant portion of the market is willing to pay a premium price (+15%) for rabbit organic farming, which is a highly valued attribute in comparison to conventional products.

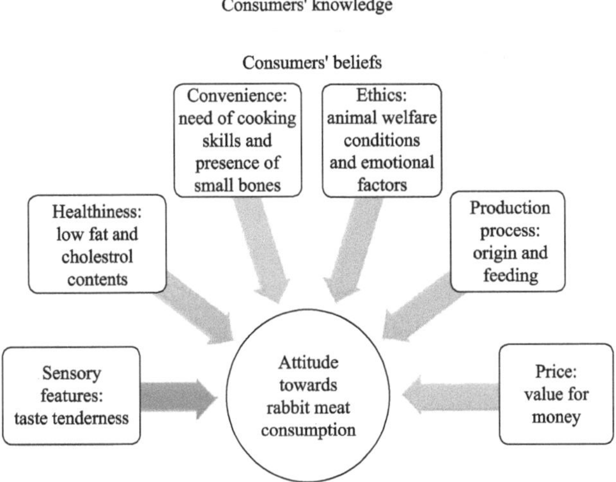

Fig. 1: Consumer views on the consumption of rabbit meat are influenced by their beliefs

As a result, the favourable perception of organic goods gives manufacturers a chance to stand out from the competition in the rabbit meat market while simultaneously promoting sustainability by preserving the environment and animal welfare. Ethical convictions are an additional factor that influences how customers feel about eating rabbit meat. Because they view rabbits as pets or rodents, a certain percentage of customers abstain from eating rabbit meat for moral and emotional reasons. Attitudes towards aspects related to the manufacturing process, such as origin and feeding, also influence attitudes towards the consumption of rabbit meat. Because of its quality and dedication to the local economy, consumers place a premium on the meat's provenance.

Consumer Attitudes Towards Functional Foods Including Rabbit Meat

Demand for functional foods has surged as a result of consumers' rising knowledge of the intimate connection between diet and health. Functional foods are foods with specific nutritional qualities that, when consumed, improve bodily processes and/or lower the risk of illness. Foods that have had one or more nutritional components added, removed, or altered using technological or biotechnological methods might be considered functional products. customers' growing interest in functional foods is a strategic opportunity for the meat business, as it may diversify its product line to increase quality and counteract customers' unfavourable opinions of meat's unhealthy qualities. Additionally, encouraging the use of functional foods is a smart way to help down the expense of public health. There are two ways to obtain functional meat: removing ingredients that have detrimental effects on health or adding ingredients that have positive effects. The addition of functional ingredients (vegetable proteins, fibres, spices, herbs, probiotics, etc.) during meat processing; (ii) improving the formation of functional compounds during processing; and (iii) adding functional nutrients (e.g., n-3 polyunsaturated fatty acids (n-3 PUFA), conjugated linoleic acid, vitamins, minerals) to the animal diet are specific strategies for improving the functional features of meat.

There are very few, and even fewer studies on consumer acceptability and desire for meat with functional qualities when it comes to rabbit meat. On the one hand, meat is viewed by consumers as a wholesome diet that is rich in protein and healthy fats like omega-3 polyunsaturated fats. On the other side, eating red meat—especially processed meat—raises your risk of death, cancer, and cardiovascular disease. For these reasons, meat with functional qualities like lower fat and salt levels is in greater demand from customers. Because it naturally has many nutritionally beneficial qualities, including a high percentage of omega-3 polyunsaturated fatty acids, protein and vitamin contents, and lower fat and cholesterol contents than other meats, rabbit meat demonstrates that it can be a useful meat type for consumer health. You may enhance the already exceptional nutritional qualities of rabbit meat by adding natural feed, such linseed, thyme, or bilberry, to the animals' meals. Experiments showed that the modified rabbit feed improved the meat's fatty acid composition (an increase in n-3 PUFA) without changing the meat's physical or sensory characteristics. Furthermore, compared to cage rearing, rabbits raised on a free-range diet had higher meat concentrations of n-3 PUFA and higher antioxidant capacities.

Although consumers want food that has a positive impact on their health, they are often unaware of foods that have these qualities, like rabbit meat. The results emphasise the value of consumer education programmes to raise consumer awareness of the various nutritional benefits and practical applications of rabbit meat. In this sense, functional meat that plays a significant part in customers' meals for disease prevention can be obtained by feeding and combining rabbit meat with other components that have functional value. Consuming meat, particularly red meat, is blamed for a number of health problems; however, eating white meat, like rabbit, may contain healthy substances. Producers that could create functional meat could have a chance to capitalise on the increasing customer demand for these kinds of goods. By meeting customer desire for food that contains components that are beneficial to health, a meat-based functional meal could enhance the perception of meat. Furthermore, encouraging the consumption of white meat that contains beneficial compounds—like rabbit meat—can be essential to lowering the cost of public health care.

Rabbit Farming Project (100+20) Cost and Profits

The report below presents a model project for rabbit farming, specifically consisting of 100 female rabbits (does) and 20 male rabbits (bucks).

Introduction of the Project

The rabbit rearing business is experiencing a surge due to the increasing demand for meat. There has been a growing interest among individuals to venture into commercial rabbit farming alongside their current livestock enterprises. Critical aspects of this type of farming project include breeding, caring for young rabbits, and managing feed for the rabbits.

1. Starting a rabbit farming project offers numerous benefits that make it a worthwhile endeavor. One advantage is that rabbits are known for their rapid growth rate, surpassing that of many other animals.
2. Additionally, rabbits possess an exceptional feed converting ratio, making them highly efficient in converting feed into valuable resources.
3. Another advantage is that running a commercial rabbit farming business does not necessitate any specialized skills, making it accessible to a wide range of individuals.
4. Multi-kidding is a common occurrence in rabbits, with a single female rabbit being able to produce anywhere from 3 to 8 offspring in a single birth.
5. Rabbit farming is a space-efficient venture, as it does not require a large amount of space.
6. The setup and production costs associated with rabbit farming are minimal, making it a cost-effective option.

7. Rabbit meat is not only lean but also highly nutritious, making it a healthy choice for consumption.
8. Rabbit meat is enjoyed and consumed by people all around the world.
9. Rabbit feed costs are relatively low, as they can be sustained on kitchen wastes, grass, and plant leaves.
10. Managing a rabbit farming project requires minimal labor, making it an efficient and manageable endeavor.
11. Rabbit farming can be easily handled by women and even older individuals, as it does not demand excessive physical strength.
12. If you decide to establish a commercial rabbit farming project, the initial investment required is relatively low, and it can be recouped within a few months.
13. In addition to meeting your family's meat needs, you can also sell rabbit manure, adding an additional source of income.
14. Commercial rabbit farming is a profitable and viable business opportunity.
15. Rabbit (Black) (meat purpose) available for Rabbit Farming Project
    - New Zealand White, Soviet Chinchilla, Grey Giant, White Giant.
16. **Rabbit Breeding Age for Rabbit Farming Project**: 6 to 8 months.
17. **Number of Rabbits Per Unit in Rabbit Farming Project**: 100 female rabbits plus 20 male rabbits.

Rabbit Breeding & Rearing Cycle in Rabbit Farming Project

Pregnant Rabbit

1. The ratio of male rabbit and female rabbits = 1:5.
2. Pregnancy duration = About 1 1month (30 days).
3. Kindling percentage = 80%(for every 100 female rabbits, 80 will be productive/pregnant).
4. Average No. of young rabbits born/kindle = 6.
5. No of kindlings in a year per rabbit = 4.
6. Female rabbits bred again = 1 week after weaning.
7. Number bunnies obtained = 80 females x 6 bunnies x 4 kindlings = 1900.
8. Mortality rate in bunnies = 30% = 560.
9. Young bunnies available in the farm = 1900 – 560 = 1340.

10. Mortality in adult rabbits = 5 to 10%.
11. Average adult rabbit body weight = 3 kg to 3.5 kg.
12. The average live weight of bunnies at 90 days old (3 months age) = 1 kg.
13. Cage size for rabbit = 360 square inches for an adult rabbit.
14. Concentrate feed required for male rabbit = 120 grams per day, female rabbits (dry, for pregnant rabbits = 120 grams per day and for weaner of 6 to 12 weeks = 50 grams per day.
15. Hay requirement in rabbit farming = Male rabbit 40 to 45 grams per day, for female rabbit (lactating) = 40 grams per day and for weaner (6 to 12 weeks) = 30 grams per day.
16. Meat yield: For young rabbit of 3 to 4 months age = 1 kg, and above 4 months = 1.5 kg.
17. Cost of rabbit meat = 120 Rs/kg.
18. Manure income from rabbit farm = 2 rupees per rabbit.

Newborn Rabbits

Spread over of breeding rabbits and their growth details in Rabbit Farming Project					
Month	**Parameter**	**1st Batch**	**2nd Batch**	**3rd Batch**	**4th Batch**
January	Breeding				
February	Kindling	Born, Growth			
March	Breeding	Growth			
April	Kindling	Growth, Sold	Born, Growth		
May			Growth		
June	Breeding		Growth, Sold		
July	Kindling			Born, Growth	
August				Growth	
September	Breeding			Growth, Sold	
October	Kindling				Born, Growth
November					Growth
December					Growth, Sold

Herd projection for a rabbit farming project unit of (100 females plus 20 males)				
	Adult Rabbits		**Young Rabbits**	
	Male rabbits (Bucks)	**Female Rabbits (Does)**	**Male rabbits (Bucks)**	**Female Rabbits (Does)**
By purchase	20	100	Nil	Nil
By Breeding			950 (From above mentioned Rabbit Breeding & Rearing Cycle of 4 kindlings)	950 (From above mentioned Rabbit Breeding & Rearing Cycle of 4 kindlings)
Total	20	100	950	950
Mortality (Death)		10	670	670
Sale of Rabbits		20	670	670
Balance at the end	20	70	0	0

Capital cost/expenditure for Rabbit Farming Project

A) **Sheds and cages:**

- Cost of construction 1shed kacha: 12 feet x 30 feet = Rs. 28,800
- Cages : (120 cages x Rs. 100) = 12,000
- Cost of construction 1shed kacha 12 feet x 30 feet = Rs. 28,800
- All wire cages: 180 x Rs. 55 = Rs. 9,900

  (Note: 11/2 feet x 11/2 feet for growers)

B) **Daily use items:**

Buckets, wire brushes, blow lamps, feeders and waterers/nestboxes = Rs. 3,500

Total A+ B = Rs. 83,000

C) **Cost of broiler rabbits:**

100 female rabbits (adults) plus 20 male rabbits (adults) @ Rs. 300 = 120 x Rs. 300 = Rs. 36,000

Total from above (A+ B+ C) = Rs. 1,19,000

Recurring cost / expenditure (12 months for initial stock and 2 months for young)

Rabbit Feed

a) Feed cost for adults: 120 x 0.12 x 365 (52.56 q x 750) = Rs. 39,420
b) Feed cost for young: 1340 x 0.05 x 60 days (40.20 x 750) = Rs. 30,150
c) Hay cost lump sum = Rs. 10,000
d) Miscellaneous expenditure = Rs. 20,000

Total from above (a + b + c + d) = Rs. 99,570

Income for the 1st year from Rabbit Farming Project

- Sale of meat 1340 kg from 1340 young of 12 weeks old @ 120/kg = Rs. 1,60,800
- Sale of 1340 rabbit skins @ Rs 80 each = Rs. 1,07,200
- Sale of manure 12 kg/per adult /yr @ Rs 2/ kg = Rs. 2,880
- Sale of manure 1340 x Rs. 1 = Rs. 1,340

Total from above = Rs 2,72,222

Cost of production in Rabbit Farming Project

- Recurring expenditure for I year = Rs. 99,570
- Depreciation @ 10% on fixed amount = Rs. 8,300

Total from above = 1,07,870

Net profit for the I year from Rabbit Farming Project

Rabbit Farming Project

Income = Rs. 2,72,222

Expenditure = Rs. 1,07,870

Profit per animal = Rs. 80.5

Net income = Rs. 1,64,352

Non-Recurring expenditure in Rabbit Farming Project = Rs. 1,19,000

Banks can provide 75% of the total cost of the rabbit farming project. The following table lists out with the loan amount and net profit in rabbit farming project report for five years duration.

Bank loan amount for Rabbit Farming Project = 75% of Rs. 1,19,000 = Rs. 89,250

Assume principle bank interest rate as : 14%

	First year (in Rs.)	**Second year (in Rs.)**	**Third year (in Rs.)**	**Fourth year (in Rs.)**	**Fifth year (in Rs.)**
Principle	89,250	71,400	53,550	35,700	17,850
Interest rate @ 14%	12,495	9,996	7,497	4,998	2,499
Loan portion	17,850	17,850	17,850	17,850	17,850
Total	30,345	27,846	25,347	22,848	20,349
Total income / year	1,64,352	1,64,352	1,64,352	1,64,352	1,64,352
Net profit in Rabbit Farming Project	**1,34,007**	**1,36,506**	**1,39,005**	**1,41,504**	**1,44,003**

10

Future Rabbit Research in India

In recent years, the demand for rabbit meat and products has been steadily increasing in India. This has led to a growing interest in rabbit farming and research in the country. The future of rabbit research in India looks promising, with ongoing efforts to improve breeding techniques, nutrition, and disease management. Some of the key areas of focus include developing high-yielding breeds that are well adapted to the Indian climate and improving feed formulations to optimize growth and productivity. As there is also a rising interest in rabbit fur production, research is being conducted on ways to enhance fur quality and increase its market value. Furthermore, studies are being carried out on the potential health benefits of consuming rabbit meat as a source of lean protein. The advancement of technology has also opened up opportunities for genetic research and the development of new vaccines against common diseases affecting rabbits. With support from government agencies, universities, and private organizations, it is expected that the future will see significant progress in rabbit research in India, leading to a more sustainable and profitable industry for farmers across the country. In this chapter, we will discuss the future prospects of rabbit research in India.

Breeding and Genetic Research

Breeding and genetic research play a vital role in enhancing the production potential of German Angora rabbits, especially in sub-temperate agro-climatic conditions. Despite their high wool yield, challenges such as importing superior germplasm and inbreeding issues persist due to continuous use of the same germplasm by farmers. To overcome these challenges, selective breeding is necessary to improve local Angora rabbit stocks. This can be achieved by implementing a multi-trait selection index and developing divergent lines in German Angora, leveraging hybrid vigor for growth, wool yield, staple length, fiber diameter, and guard hair. Additionally, efforts should focus on reducing the shearing interval, currently at 75 days or more, to increase annual wool yield and enhance wool quality, benefiting both breeders and the Angora rabbit industry as a whole.

Selecting superior males from farmers' flocks based on their performance under uniform conditions and distributing them among farmers with lower-grade animals is essential for accelerating genetic improvement and preventing inbreeding. Establishing a central germplasm center with regional and sub-centers is crucial for providing high-quality males to farmers as needed, ensuring the use of the best breeding stock. Regular supply of breeding stock from research units to these regional germplasm resource centers is necessary to maintain stock quality. The management of these centers should be entrusted to government agencies, non-governmental organizations, or progressive farmers trained in effective breeding programs and record-keeping. As research progresses, identifying highly prolific animals through genetic marker studies and utilizing marker-assisted selection to identify superior parents for the next generation will further advance livestock genetics and contribute to sustainable agriculture.

Meat Science Research

Meat science research has traditionally focused on measuring factors such as pre-slaughter age, carcass weight, dressing percentage, and carcass characteristics. While this information is valuable for understanding the quality of meat, there is a lack of studies exploring the composition of meat and how it can be utilized in different products. Further research is needed to develop easy-to-cook and consumer-friendly meat products.

In addition, there is a need for extension studies that promote the benefits of rabbit meat over other meats such as chevon, mutton, veal, pork, and chicken. These efforts can be made through mass media channels like radio, TV, and newspapers to reach a wider audience. It is important to educate consumers about the nutritional value and cost-effectiveness of rabbit meat compared to other meats.

Consumers are increasingly concerned about the health aspects of their food choices. Therefore, future research should prioritize highlighting the low sodium, cholesterol, and fat content of rabbit meat. Additionally, researchers should emphasize its high levels of polyunsaturated fatty acids which make it a healthy choice for consumers.

Overall, there is still much to be explored in the field of meat science. By expanding our knowledge on factors such as composition estimation and product preparation techniques, we can better cater to modern consumer demands for nutritious and convenient meat options. Through continued research efforts and effective communication with consumers, we can promote rabbit meat as a sustainable and healthy protein source in today's market.

Wool Science Research

Extensive research on Angora wool has been conducted and documented in recent literature, however, the main objectives have not been achieved. Due to price fluctuations in the local market, many Angora farmers have been forced to abandon their profitable business. Therefore, there is a pressing need for marketing research to establish a fixed price for wool regardless of seasonality, which will enable farmers to resume and sustain their ventures profitably. Furthermore, most locally produced wool is evaluated in other regions, emphasizing the need for local facilities for testing and processing. It is crucial to strengthen extension research through mass media channels to promote the benefits of Angora farming and provide support to discouraged farmers through developmental agencies. Exploring opportunities for value addition of Angora wool by blending it with other natural and synthetic fibers and addressing issues such as hair shedding should also be given further consideration in future research endeavours.

Reproduction Research

Rabbits should be bred when they reach 80-85% of their adult body weight, as they are capable of nursing their offspring at this stage. Therefore, reproduction studies should focus on breeding does as soon as they reach their optimal body weight. Approximately 60-70% of ovulated eggs are fertilized, with 30-40% remaining unfertilized or lost during early stages of implantation. To maximize fertilization and conception rates, it is essential to provide suitable environmental conditions and mitigate factors that contribute to these losses.

Efforts should be directed towards minimizing embryonic losses to increase production by maximizing the number of kits per doe during kindling. Pseudopregnancy, a common occurrence in rabbits, results in a two-week delay in normal cyclicity for does, wasting the breeding potential of superior animals. Addressing this issue in routine management practices can improve kindling percentages in different climates.

Future research should focus on reducing pseudopregnancy occurrences to enhance production. Additionally, exploring methods to increase ovulation rates and births per kindling through hormonal interventions, such as gonadotrophins, can be beneficial. While the use of eCG in rabbit reproduction is controversial due to anti-eCG antibody formation, alternative bio-stimulation methods recommended by the International Rabbit Reproduction Group (IRRG) can enhance doe receptivity post-kindling.

Recent studies have shown that indigenous plants can induce estrus in anoestrus rats and mice, leading to multiple ovulations. Investigating plants with gonadotropic properties for hyperstimulation of gonads in rabbits could be a promising avenue for future research.

Health Management Research

Health management research is essential for understanding and enhancing the well-being of individuals and communities. However, there is a lack of reporting on mortality patterns of diseases in various zones, hindering the identification of prevalent diseases and their outcomes. A nationwide survey is necessary to determine the susceptibility of different areas to specific diseases. Research is also needed to develop targeted treatments for conditions like Tyzzer's disease, mucoid enteritis, pneumo-enteritis, bloat, rhinitis, car-kanker, wry neck, and hairball through experimental trials. Additionally, a health calendar should be created based on disease prevalence in different seasons and geo-agroclimatic regions. Intensified research on the prevalence, prevention, and treatment of diseases affecting kit and adult survivability in farmers' flocks is crucial for effective health management. Comprehensive research in these areas can provide valuable insights for improving health outcomes and developing interventions tailored to different disease risks in various zones. Systematic studies are necessary to investigate deadly diseases impacting production through mortality and morbidity, including factors contributing to conditions like wry neck, hind leg paralysis, and diseases such as coccidiosis and enteritis for future prevention programs.

Nutrition Research

Research should focus on developing cost-effective feeding systems for Angora rabbits during periods of scarcity and optimizing wool production during high market demand. Maize, a readily available energy source in hilly regions, is suitable for feeding. However, the lack of affordable protein sources is a key area for investigation. Evaluating local, low-cost protein sources can help reduce feed expenses and ease the financial burden on farmers during periods of low wool prices. Standardizing nutritional requirements for different life stages of Angora rabbits (weaners, young, grower, adults, pregnant/lactating, and breeding animals) is essential.

Additionally, research should explore the use of locally available agro-industrial by-products to formulate cost-effective rations for Angora rabbits. Evaluating the effectiveness of these formulations in terms of growth, production, and reproductive performance is crucial for ensuring profitable Angora farming. Furthermore, studies on the impact of nutrition on wool quality in Angora rabbits and carcass composition in broiler rabbits are needed to enhance overall productivity and profitability in rabbit farming.

Physiology and Climatology

To enhance wool production in breeding animals, it is important to maintain a small breeding flock through an intensive system. This approach maximizes offspring production without compromising overall flock productivity. The climate significantly impacts Angora rabbit production, with the sub-Himalayan temperate region being ideal. Areas above 1500m altitude from Jammu and Kashmir to Arunachal Pradesh are highly suitable, as well as the Nilgiri hills.

Angora rabbits thrive in temperatures of 2-35°C, with their comfort zone around 15-18°C. Maintaining humidity at 60-65% is crucial for their health. Proper lighting is essential for maximizing wool production. Providing 30-50 lux of light for 14-16 hours daily promotes healthy growth in Angora rabbits.

By managing environmental factors like climate, lighting, and breeding animal numbers, farmers can achieve peak wool production. Monitoring and controlling obnoxious gases in rabbit sheds, such as keeping ammonia below 30 ppm, carbon dioxide under 300 ppm, and sulfur dioxide under 10 ppm, is vital for rabbit health. Ensuring access to drinking water and green grass within cages is essential for rabbit welfare. Ongoing research in environmental physiology is crucial for establishing optimal conditions for Angora rabbit production and maintaining high standards of animal care.

In conclusion, maintaining appropriate gas levels and providing essential resources are critical for rabbit health. Continued research on improving environmental conditions for Angora rabbit production is essential for advancing responsible animal husbandry practices.

Social Science Studies

Numerous social science studies have shown that rabbits have a gentle and friendly nature, making them easy to handle for people of all ages. This quality makes rabbit farming a great option for individuals looking to earn a livelihood without disrupting their daily routines. Women, children, and individuals with disabilities can easily participate in rabbit farming due to the low level of effort required. It is important to consider the role of women in rabbit farming and their level of involvement. Further research is needed to explore the potential of rabbit farming as a professional career choice for educated youth, a rehabilitation program for ex-service personnel, and an income-generating opportunity for physically challenged individuals. A more comprehensive investigation into these possibilities is necessary to fully understand the benefits that rabbit farming can offer to different individuals and groups in our society. In conclusion, the friendly nature of rabbits makes them a practical and inclusive option for income generation while catering to the diverse needs of our community.

Organic Rabbit Farming

Organic farming has emerged as a global phenomenon, capturing the attention of consumers, policymakers, and farmers alike. With an increasing emphasis on food quality and safety, organic production has become a popular choice for those seeking healthier and more sustainable food options. In many developing countries, achieving food security is a top priority. However, there is also a growing need to establish specialized centers for organic production - known as "Growth Centers" - to meet the rising demand for high-quality organic products. These centers would serve as hubs for research, development, and training in organic farming techniques.

Prime locations for establishing Growth Centers include hilly regions, forested areas, and rain-fed regions with underdeveloped agriculture. These areas have untapped potential for organic farming and can benefit greatly from such initiatives. One promising venture that could be implemented in all parts of the country is broiler rabbit farming. Often overlooked compared to other livestock options, rabbits are efficient converters of feed into meat and can thrive in various environments. Their low carbon footprint also makes them an environmentally-friendly choice.

Broiler rabbit farming not only offers economic opportunities but also aligns perfectly with the principles of organic farming. By raising rabbits without hormones or antibiotics and allowing them to graze on natural vegetation sources, their meat becomes free from harmful chemicals and additives. The establishment of a Growth Center for organic rabbit farming could have significant positive impacts on both local economies and the overall organic food market. It would provide employment opportunities for rural communities while promoting sustainable agriculture practices. Moreover, the addition of high-quality organic white meat to the market will cater to the growing demand for healthier protein sources among health-conscious consumers. This would further contribute to the growth of the organic industry globally.

In conclusion, establishing a Growth Center for organic rabbit farming in suitable areas across our country has immense potential for both economic growth and promoting sustainable agriculture practices. Further research and development in this area will not only benefit local communities but also contribute to the global movement towards a more sustainable and organic food system.

Index

A

Abscesses 104-107, 113
Altex Rabbit 26
Amino acids 119
Angora 1, 5, 13-19, 24, 87, 131, 133, 134, 135,
Antioxidant 124
Artificial insemination 91-92, 95-96
Automatic Waterer 49

B

Bamboo Hutch 40
Bamboo Troughs 43-44, 48
Beveren 29, 31
Blanc De Bouscat 24
Body weight 2, 3, 14, 15, 18, 60, 61, 94, 98, 100, 127, 133
Breeding and Genetic Research 131
British Angora 5, 14
British Giant 25

C

Caecal impaction 109
Caecotrophy 4
Cage system 50-52
Cage system 50-52
Cans 46-47, 50, 88
Capital cost 128
Carbohydrates 62
Ceramic Crocks 49
Checkered Giant 27
Climate 7, 12, 18, 20, 35, 50, 92, 100-101, 131, 133, 135
Coccidiosis 6, 9, 51, 67, 85, 96, 107-108, 116, 134
Colony housing system 53-54
Concentrates 7, 61, 63-65
Continental Giant 22-23
Coprophagy 67
Counterset Nest Box 57
Crocks 42, 47-49, 83
Crossbred Angora 18
Cutaneous staphylococcosis 104

D

Disease Control 83, 103
Doe-litter separation 98
Dutch 29, 33

E

E. coli infections 106
Enamel Cups 47
Encephalitozoonosis 112
English Spot 29, 32-33
External parasites 109

F

Fancy 11, 29
Fiber 13, 15-16, 18-18, 61-63, 87, 133
Flemish Giant 11, 16, 20, 21, 22, 24, 26, 32
French Angora 15-17, 24

G

German Angora 1, 5, 12, 14, 16, 19, 131
Giant Angora 16-17
Giant Chinchilla 25-26
Giant French Lop 22-23
Giant Papillon 28
Giant 2, 5, 11, 16-18, 20-28, 32, 91, 126
Grass Mangers 43
Gray Giant 2, 5, 28

H

Hairballs 110
Hanging cage system 51-52
Havana 29-31
Health Management Research 134
Hoppers 44, 46
Humidity 7, 89, 101, 135
Hungarian Giant: 24-25
Hutch system 36, 52-53, 67

K

Kindling 55-56, 68, 70-71, 74-77, 81, 83, 96, 98, 115, 126-128, 133

L

Lactation 61, 68, 97
Lighting 3, 76, 88, 91, 97-98, 100, 111, 119, 132, 135

M

Marketing 8, 69, 117, 120-121, 133
Mating 53, 70, 91, 93-95
Meat Production 3-4, 6, 8, 13, 20-21, 63, 117
Meat Science Research 132
Moist dermatitis 113
Myxomatosis 9, 110-111, 115

N

Nest Boxes 55, 57, 83, 96
Net income 129
Net profit 129
New Zealand Red 29, 32
New Zealand White 2, 5, 20, 21, 26, 92, 93, 126
Non-Recurring expenditure 129
Nutrition Research 134

O

Organic Rabbit Farming 136

P

Palomino 29-30
Palpation 73-75, 104, 110
Physiology and Climatology 135
Polish 11, 29-30

R

Rabbit Farming Project 125-129
Rabbit tractor cages 54-55
Record Keeping 79
Recurring cost 128
Red urine 114
Religious taboo 6
Reproduction Research 133
Ring worm 112
Russian Angora 5, 15

S

Satin Angora 17
Silver Fox 20, 27
Snuffles 51, 85, 103-104, 107, 115
Social science studies 135
Sore hock 108-109
Soviet Chinchilla 2, 5, 20, 126,
Spanish Giant 23
Standard Nest Box 55-56

Suckling mortality 99
Syphilis 113

T

Temperature 7, 49, 70, 89-90, 100, 104, 115, 118, 135
Toxoplasmosis 112
Tyzzer's diseases 113

V

Ventilation 35, 38, 83-85, 89-90, 101, 103-104, 114-115

W

Weaning 70-71, 77-78, 81, 93, 95-96, 98-99, 126
Wire Hutches 35, 38-39
Wool block 110-111
Wool production 12-15, 17, 134-135
Wool science research 133
Wry neck condition 107